Appropriating the Weather

Plate 1. Vilhelm Bjerknes (1862–1951). Courtesy of the
Norwegian Academy of Science and Letters.

APPROPRIATING
THE WEATHER

Vilhelm Bjerknes and the
Construction of a Modern Meteorology

Robert Marc Friedman

Published in cooperation with
Uppsala Studies in History of Science

Cornell University Press

Ithaca and London

First published 1989 by Cornell University Press.

International Standard Book Number 0-8014-2062-8
Library of Congress Catalog Card Number 88-47729
Printed in the United States of America
Librarians: Library of Congress cataloging information
appears on the last page of the book.

The paper in this book is acid-free and meets the guidelines for
permanence and durability of the Committee on Production Guidelines
for Book Longevity of the Council on Library Resources.

To my parents and brother Joel,
who first showed the way . . .

Contents

Illustrations

[ix]

PLATES

Preface

ALREADY I hear grumbling and questioning. The title. What an odd formulation. How might phenomena so fleeting, so erratic, so ubiquitous as weather be appropriated? And can knowledge of nature be constructed rather than, say, discovered or logically deduced? Why the indefinite form: "a" modern meteorology? To make matters more puzzling—and interesting—I confess that the title of the book should include: "in the age of powered flight." For Vilhelm Bjerknes, the primary actor, must share the spotlight not only with his assistants but also with some ungainly objects: airships, airplanes, and balloons both large and tiny. Several additional artifacts joined them at various times on the stage of science history: motorized fishing boats, wireless telegraphic equipment, and haystacks. These props call to mind that in the period under investigation, 1898–1925, the social significance of weather—and with that its scientific import—changed remarkably. Weather's increased role in the political economy of nations was not simply a consequence of Bjerknes's science but a constituent in the making of it.

During the first quarter of the twentieth century, meteorological theory and practice underwent a major transformation. Important breakthroughs occurred especially between 1918 and 1925, when a group of Scandinavian researchers under the leadership of Vilhelm Bjerknes (1862–1951) established a new conceptual foundation for atmospheric science. The formulation of the so-called Bergen meteorology, based on the concepts of fronts and air masses, has long been recognized as one of the principal turning points of this science. The achievements of the Bjerknes group include a new model of the ex-

tratropical cyclone, that is, the type of low-pressure system common
in the midlatitudes. Claiming that these atmospheric disturbances are
composed of three-dimensional surfaces of discontinuity—fronts—
the scientists working in Bergen began to conceive the cyclone as a
wave that develops and grows along preexisting polar fronts that
separate polar and tropical air masses. In shifting from regarding
cyclones as undifferentiated entities to regarding them as composed
of fronts, they were eventually able to provide the first clear physical
explanation of cyclone evolution. Bjerknes's endeavors marked a decisive turn to a dynamic-physical comprehension of the atmosphere
from the statistical-climatological understanding dominant since the
late nineteenth century. As a consequence the Bergen meteorology
promoted an unprecedented interaction between theory and practice,
which facilitated the rapid growth of meteorology as a scientific discipline during the next quarter century. The emergence of this new
atmospheric science was a complex development, for which both
Bjerknes's career goals and the increased political economic significance of weather proved crucial.

Knowledge is power: the familiar Baconian adage takes on further
significance in modern science and is central to any full understanding of Bjerknes's career. Bjerknes grasped the social processes of
science. As conditions affecting his research programs changed, he
took adaptive measures to continue advancing his disciplinary aims.
Although the term *discipline* is generally used to refer to the institutional, personnel, and cognitive resources of a science, a more general
meaning of the word is pertinent here: a means of instilling or reg-
ulating a set of rules, conduct, or behavior. Scientific disciplines are
also social systems in which members pursue, negotiate for, and compete for authority and resources with which to participate in, if not
direct, the production and circulation of knowledge. This concept of
a scientific discipline yields the central theme of this book: appropriating the weather.

Appropriating—taking to oneself as one's own, setting apart for a
purpose—provides the key to the interconnected history of Bjerknes's career and the construction of a new meteorology. Diverse
forms of appropriation of weather run through this history. In the
early 1900s, Bjerknes turned to atmospheric studies to avoid professional decline as a theoretical physicist. He hoped to remake meteorology into a mechanical physics of the atmosphere: weather was to
be incorporated exclusively into the domain of physics. But by the
end of World War I, Bjerknes recognized that to continue striving
toward this goal he would also have to appropriate the weather for
commerce. By providing detailed, precise forecasts based on physical

principles that might enable aviators, farmers, and fishermen to use the weather for efficient operations, Bjerknes could nurture a school of scientifically trained meteorologists. Believing that the system of concepts and techniques they were developing in Bergen could best satisfy the problems facing postwar meteorology, he and his assistants campaigned to spread their forecasting practices and research agenda abroad. The Bjerknes group's endeavors to have its science dominate international meteorology was yet another appropriation of the weather. In the broadest sense Bjerknes's enterprises might be understood as appropriating the weather for himself and his assistants, appropriating it as professional property that could be exchanged for authority, resources, and prestige in the world of science and in society.

"To appropriate" implies an object to be appropriated that already exists, waiting to be set aside and possessed; yet weather of course is too ethereal for any such direct subjugation. Actually, weather was redefined, reclassified, and restructured. First, Bjerknes abstracted weather as a problem in hydrodynamics. Then, in Bergen, as he and his colleagues were devising innovative forecasting practices, they constituted three-dimensional atmospheric concepts and integrated them into meteorology. Their new concepts and models were neither "discovered"—there to be seen in the observations—nor devised during theoretical inquiry. Theory and empirical data did contribute to the shaping of these constructs but did not and could not lead to them or define them exclusively. It was changes in the goals, methods, and technological basis of forecasting that led to the new concepts that became the foundation for meteorology.

This is a complex chapter in the development of modern atmospheric science. Early on I recognized that any attempt to make sense of the events by restricting my inquiry to a set of ideas isolated from the broader historical context was doomed to failure. Moreover, assembling a list of so-called precursors who allegedly almost saw and said that which is traditionally ascribed to the Bergen school illuminates little and explains nothing. History of science should not be an arena for awarding points for discovery and innovativeness; at its best it offers insight into the nature of science, its epistemological claims and its complex interactions with society and culture.

The recent past, seemingly such familiar territory, can deceive. Attempts to comprehend the history of early twentieth-century meteorology using today's understanding of this discipline lead to confusion. Neither the present sophisticated theories and practices nor contemporary meteorology's highly developed institutional and organizational infrastructure, full academic legitimacy, and division of scientific labor can provide useful categories for analysis. Bjerknes worked in another era. When I began interpreting the enormous body of relevant unpublished and published documents, I sought to recover what Bjerknes and his assistants were "doing" (as Quentin Skinner might ask) and what they expected to accomplish by their actions. Rarely did events develop according to Bjerknes's expectations. I found it necessary to reestablish contexts of meaning and intention to interpret the documents and to make sense of Bjerknes's career.

In spite of the complex themes involved, I have chosen to present this history through a narrative analysis unfolding over time. I hope the book will be of interest and value to historians, philosophers, and sociologists of science, to meteorologists and other geophysical scientists, and to persons interested in the history of Norwegian and Swedish science. Naturally, to address a diverse audience, I was forced to make compromises in detail and content. Although biographical detail belongs to the narrative, the book does not purport to be a biography of Bjerknes; nor does it pretend to be a comprehensive history of the meteorology of this period, or even of the Bjerknes school's endeavors. I hope my analysis of central problems and themes is more interesting and significant than an exhaustive compilation of factual material through which few readers would care to wade.

Finally, in researching and writing this study, I developed perspectives and interpretations that conflict with some received ideas on the history of this particular chapter of meteorology and on the nature of the scientific enterprise. Still, I do not consider this book a case illustration of any particular model or theory of scientific change. Readers may discern the ghost of the late Michel Foucault lurking in some analyses and questions; yes, but I would hardly consider this work an application or test of his theses or methods. Issues do arise in the narrative related to—among other themes—the interaction of practice and theory, the dynamics of disciplinary development, the pursuit of science in liberal-economic societies, and the role of utilitarian concerns in problem choice and theorizing. But rather than appealing to an ever-growing literature in the social studies of science for understanding, I seek it first in the details of why and how such issues arose and then assumed particular configurations in a specific time and context. Although I did not write this book to be polemical, I hope it will stimulate discussion.

For over a decade I have returned on several occasions to the historical study of Vilhelm Bjerknes and his endeavors; however, Bjerknes and the Bergen school have been familiar names to me for most of my life. As a boy, I discovered for myself the fascination of weather; in the books read by youthful amateur "weathermen" I soon met that exotic and seemingly unpronounceable name from Norway. Subsequently, I received a formal education in geophysical science at what was then New York University's Department of Meteorology and Oceanography. My decision not to continue in this field was difficult; like most first loves, meteorology still elicits within me emotions and regrets.

When I enrolled in the Department of History of Science at Johns Hopkins University, I did not plan on studying the history of atmospheric science. For my eventual decision to write my dissertation in this area, I am indebted to Professor of Meteorology George S. Benton. Over the years I have benefited both from his encouragement and from the many insights he offered me during our discussions. A summer research grant from the Advanced Study Program at the National Center for Atmospheric Research proved equally instrumental in promoting my interest in history of meteorology. When I finally decided on the Bergen school of meteorology as a dissertation topic, I received generous assistance from the American-Scandinavian Foundation, the Fulbright Exchange Program, the National Air and Space Museum of the Smithsonian Institution, the National Science Foundation, and the Norwegian Research Council for Science and the Humanities. While in Norway for the first time, in 1975 and 1976, I enjoyed the hospitality of Professor Arnt Eliassen and his staff at Oslo University's Institute for Geophysics. When I returned to Norway a few years later to revise the dissertation, I had no idea that I would become so deeply immersed in the subject and remain for so many years in Scandinavia.

I completed the research and a preliminary draft for this book by 1983; other engagements interrupted the writing several times. While working on this book I received assistance, guidance, and inspiration from many persons and institutions, for which I am deeply appreciative.

My debt to the Department of History of Science at Johns Hopkins University is great. There I acquired my understanding of history of science. Only after leaving could I appreciate how much I learned from my fellow students and from faculty members William Coleman, Owen Hannaway, Donna Haraway, Robert H. Kargon, Camille Limoges, Russell McCormmach, and Harry Woolf. To my friends Stephen J. Cross and Michael I. Freedman, who never ceased to harass me with questions and insights, I am grateful; for without our many conversations, whatever merit this book has would be diminished. Most of all, I thank Professor Robert H. Kargon, who as teacher, scholar, and friend has contributed most to my development as a professional historian of science. His enthusiasm for the subject and his personal concern for colleagues and students alike have been an inspiration.

The Norwegian Research Council for Science and the Humanities supported my investigations for slightly over three years from its now–long-defunct science studies program (Utvalg for viten-skapsteori). I hope this book stimulates additional studies in the history of Norwegian science, a subject that deserves greater attention. The research council's Institute for Studies on Research and Higher Education (NAVFs utredningsinstitutt) has been my home in Norway for many years; the hospitality afforded me by Director Sigmund Vangsnes and his staff has been truly generous. I thank Inger Henaug for typing earlier drafts of this book: if only we all had her patience! And I especially thank Head of Department Hans Skoie, who first invited me to the institute. I much appreciate his unfailing support of my work and his friendship.

Finally, the last revision of the manuscript was made across the border, and perhaps rightly so. Although Bjerknes was Norwegian, the establishment of the new meteorology is an achievement for which Sweden too has a right to claim some honor—as distasteful as this may be to some national purists. Professor Tore Frängsmyr and the members of the Office of History of Science at Uppsala University have shown that Swedish hospitality—just like Norwegian—most certainly is a national resource, abundant, of high quality, and always prized. Their patience and support have been most helpful.

Many libraries and archives provided invaluable service by placing materials at my disposal and helping me locate documents. For this help, and for permission where necessary to quote from these materials, I thank the following institutions and their staff: the manuscript divisions of the American Institute of Physics's Niels Bohr Center for History of Physics (New York), Deutsches Museum (Munich), Royal Swedish Academy of Sciences (Stockholm), Oslo University Library, Stockholm University Library, Uppsala University Library, Graz University Library, and Karl Marx University Library (Leipzig); the National Archives in The Hague, Helsinki, Oslo, Paris, and Stockholm; the Regional State Archives in Bergen and Kristiansand; the libraries of the Norwegian Meteorological Institute (Oslo), Norwegian Museum of Technology (Oslo), West Norway Weather Bureau (Vervarslinga på Vestlandet) (Bergen), Uppsala University's Meteorological Institution, Swedish Technical Museum (Stockholm), National Air and Space Museum (Washington, D.C.), and The Johns Hopkins University (Baltimore).

In addition, I am indebted to Bengt Hjelmqvist (Åryd), Charlotte Schönbeck (Heidelberg), and Einar Sæland (Oslo) for providing me with copies of Bjerknes correspondence held privately. I also thank Bengt Hjelmqvist and family for their hospitality, so many years ago. Ralph Jewell of the Bergen University graciously assisted me in locating documents in Bergen. I have also benefited from his efforts at finding and preserving documents from the early years of the Bergen school (now available at the Regional State Archive in Bergen).

For permission to quote from Vilhelm and Jacob Bjerknes's correspondence, I thank Vilhelm Bjerknes of Bergen, who acted on behalf of Hedwig Bjerknes. I also thank the editor of *Isis* and the History of Science Society for permission to reproduce in Chapter 9 portions of "Constituting the Polar Front, 1919–1920," which appeared in 1982.

Interviews with several meteorologists provided both factual material and an opportunity to meet some of the grand old gentlemen of the field. I feel honored to have met and have spoken with Tor Bergeron (Uppsala, 1976 and 1978), Jacob Bjerknes (Los Angeles, 1972), Olaf Devik (Oslo, 1975), Bernhard Haurwitz (Boulder, 1972), Jørgen Holmboe (Los Angeles, 1972), Eric Kraus (Los Angeles, 1972), and Francis W. Reichelderfer (Washington, D.C., 1978). For their comments, suggestions and questions, at one time or another, I thank Finn Aaserud, John Peter Collett, Elisabeth Crawford, Michael Aaron Dennis, Jon Elster, Michael Farley, Urban Jonsson, Matthias Kaiser, Werner Christie Mathisen, Einar Petterson, Nils Roll-Hansen, John Servos, and Brent Work.

During the long process of transforming the manuscript into a book, I have been lucky to have John G. Ackerman as my editor. He has shown patience and understanding; an editor, I have learned, can also be a gentleman. Janet S. Mais's copyediting markedly improved the text; Roger Haydon's efforts at Cornell University Press have also been appreciated.

Translations from French, German, Norwegian, and Swedish are my own; colleagues and friends have checked the accuracy of these.

ROBERT MARC FRIEDMAN

Uppsala and Oslo

Abbreviations

SAP   Svante Arrhenius Papers, KVA-SUB
SSP   Sem Sæland Papers, UBO
TBP   Tor Bergeron Papers, Meteorologiska institutionen
                (Meteorological Institute), Uppsala universitet (University)
UBO   Universitetsbiblioteket, Oslo (University Library)
UUB   Uppsala Universitetsbibliotek (University Library)
VBP   Vilhelm Bjerknes Papers, UBO
VpVA   Vervarslinga på Vestlandet (West Norway Weather Bureau),
                archives, Bergen
VWE   Vagn Walfrid Ekman Papers (private)

LITERATURE

Ab   *Aftonbladet*
AS   *Annals of Science*
AMAF   *Arkiv för matematik, astronomi och fysik*
AMN   *Archiv for mathematik og naturvidenskab*
BPfA   *Beiträge zur Physik der freie Atmosphäre*
CIWY   *Carnegie Institution of Washington Yearbook* (Washington, D.C.,
                1906–   )
DN   *Dagens nyheter*
DSB   *Dictionary of Scientific Biography* (New York, 1970–   )
DZL   *Deutsche Zeitschrift für Luftschiffahrt*
GP   *Geofysiske publikasjoner*
MWR   *Monthly Weather Review*
MZ   *Meteorologische Zeitschrift*
NMIÅ   Det norske meteorologiske institutt, Årsberetning (1918–25),
                ed. Theodor Hesselberg
Q JRMS   *Quarterly Journal of the Royal Meteorological Society*
SD   *Stockholms dagblad*
SGW   *Sächsischen Akademie der Wissenschaften, mathematisch-physischen
                Klasse*
VAH   *Kungliga Vetenskapsakademiens handlingar*
VAÅ   *Kungliga Vetenskapsakademiens årsbok*
VAÖ   *Öfversigt af Kungliga Vetenskapsakademiens förhandlingar*
VGL   *Veröffentlichungen des Geophysikalische Instituts der Universität
                Leipzig*, series 2, Spezialarbeiten aus dem Geophysikalischen
                Institut

Appropriating the Weather

Introduction

ON 22 October 1921, a storm moved past Scandinavia toward Russia. In Copenhagen the forecaster on duty at the Danish Meteorological Institute ordered a lowering of warning signals on the coast, expecting the remaining gales to subside during the next several hours. On his weather chart he noted to the west, over Britain, a weak depression of low pressure, nothing with which to be concerned, for he assumed that it would move eastward uneventfully. When the forecaster came on duty the next day in calm weather, he had no idea of the hurricane-force winds lashing the northern coasts of Denmark; communication links had been disrupted by the intense, concentrated storm. He also had no idea that ships and fishing boats were being devastated even as he continued to forecast pleasant weather. "Scandal," declared the newspapers when the damage and loss of life became known. The head of the meteorological institute, the respected theoretician Carl Ryder, explained that this storm was of a kind that could not be predicted given the present state of meteorological knowledge.

When it became known that forecasters in Norway and Sweden had correctly predicted the movement and rapid intensification of the storm, the uproar increased. All the meteorologists responsible for the accurate forecasts had been trained in Bergen; they used concepts and methods of analysis developed under Vilhelm Bjerknes's leadership. A Danish parliamentary commission ordered Ryder to visit Bjerknes and learn the Bergen system. Ryder went to Bergen, observed, and returned home; he refused to accept the Norwegian-Swedish meteorological system. He knew Bjerknes was campaigning to spread his school's work around the world; it was, after all, Bjerknes who informed the Danish inquiry that his disciples had predicted the storm correctly and who suggested that the Danes accept their innovations.[1] Bjerknes considered the endeavors of his Bergen school a step toward establishing a unified meteorological theory and practice based on physics; he intended to transform international meteorology into a rigorous scientific discipline.

Danish meteorologists, like most of their colleagues around the world, were aware that the forecasting methods at their disposal were not entirely satisfactory. These methods dated mainly from the second half of the nineteenth century, when meteorologists were optimistic that forecasting based on scientific principles and using the barometer and the telegraph might be attainable. After the invention of the barometer in the seventeenth century, natural philosophers came to associate changes in weather with fluctuations of atmospheric pressure reflected by the rise and fall of the mercury glass. During the eighteenth century, curiosity about the relations between epidemics and "airs," geographic theories of culture and behavior, and efforts to extend experimental-quantitative natural philosophy to atmospheric phenomena all whetted interest in keeping records of the changes in local weather. Several investigators attempted around 1800, by collecting weather records from learned academies around Europe, to map out the geographic distribution of weather elements at specific times in so-called synoptic charts. On 14 November 1854 a storm wreaked havoc on the British and French fleets assembled in the Black Sea during the Crimean War. A French inquiry revealed that the same storm had passed over parts of western and central Europe in the preceding days, suggesting that telegraphic exchanges of weather data between European or North American cities could be used to locate storms and perhaps to predict their movement. Pioneers such as Cleveland Abbe, Christoph Buys Ballot, Robert Fitzroy, Urbain Leverrier, Henrik Mohn, and James Smithson were soon involved in setting up national weather services.[2]

1. Discussion based on V. Bjerknes to Martin Knudsen, 4 November 1921, copy, VBP; Carl Ryder to H. H. Hildebrandsson, 28 July 1922, HHH; Danish Meteorological Institute, *Meteorologisk institut gennem hundred år 1872–1972* (n.p., n.d.), pp. 38–39, 62–63; Tor Bergeron, "A Meteorological Adventure," 1972, TBP.

2. My account here is merely a simplified background sketch; for further details, see, for example, Jim Burton, "Robert FitzRoy and the Early History of the Meteorological Office," *British Journal for the History of Science* 19 (1986), 147–76; J. L. Davis, "Weather Forecasting and the Development of Meteorological Theory at the Paris Observatory, 1853–78," *AS* 41 (1984), 359–82; T. S. Feldman, "The History of Meteorology, 1750–1800: A Study of the Quantification of Experimental Physics" (Ph.D. diss., Univ. of California at Berkeley, 1983); Jim Fleming, "Meteorology in America, 1814–1874:

Activities related to the study of weather in the nineteenth century cannot easily be summarized; theories and models of atmospheric phenomena proliferated.[3] Several assumptions and methods were common, however, to the organized nineteenth-century forecasting efforts. To identify inclement weather and storms, long associated with low pressure, forecasters drew lines of equal atmospheric pressure (isobars) on weather maps. Meteorologists not only identified, but in practice defined, the weather system carrying unsettled conditions in the midlatitudes, the extratropical cyclone, by its pattern in the pressure field: the center of the storm is at the innermost oval or curve of low pressure. The forecaster learned to infer from experience how a weather system, defined by the pressure field near the earth's surface, would move and change character. Using experience and statistics, he could also ascribe weather phenomena to the isobars' different geometric configurations.[4]

Use of telegraphy determined how much information was disseminated and how swiftly. The International Meteorological Committee, founded in 1873, helped standardize and synchronize international exchanges of observations. The predictions were neither detailed nor very specific as to time and location. They amounted to storm warnings or general weather forecasts (e.g., dry, changeable, wet) for broad regions twenty-four hours in advance. Those who needed daily predictions could follow them in newspapers, placards on important buildings, and signals consisting of flags and lanterns. The primary goal of forecasting services was to reduce the number of shipwrecks. One hundred ships were lost off the British coast in a single week in 1881; over six hundred persons drowned during another week that same year.[5] Even general warnings of gales and storms were of value to coastal shipping.

Theoretical, Observational, and Institutional Horizons" (Ph.D. diss., Princeton Univ., 1988); C. C. Hannaway, "The Société Royale de Médecine and Epidemics in the Ancien Régime," *Bulletin of the History of Medicine* 46 (1972), 257–73; W. E. K. Middleton, *Invention of the Meteorological Instruments* (Baltimore, 1969).

3. Elisabeth Garber, "Thermodynamics and Meteorology," *AS* 33 (1967), 51–65; Gisela Kutzbach, *The Thermal Theory of Cyclones: A History of Meteorological Thought in the Nineteenth Century* (Boston, 1979); W. E. K. Middleton, *A History of Theories of Rain* (London, 1965).

4. William Napier Shaw, *Forecasting Weather* (London, 1911), p. 31 and chaps. 2–5: "It is the business of the forecaster to determine what type of barometric distribution is to be expected within the next twenty-four hours, and to assign to it its appropriate weather" (p. 98); for a typical discussion of the state of forecasting, see also, A. J. Henry, "Weather Forecasting, Preliminary Statement," in U.S. Dept. of Agriculture, Weather Bureau, *Weather Forecasting in the United States* (Washington, D.C., 1916).

5. F. H. Ludlum, *The Cyclone Problem: A History of Models of the Cyclonic Storm* (London, 1966), p. 27; see Shaw, *Forecasting Weather,* for specific disasters, esp. chap. 9.

But the dream of finding simple laws for predicting weather faded by the end of the century. After three decades of studying the progression of weather systems, meteorologists had reaped a meager harvest. Many theories had been advanced; thermodynamics and hydrodynamics had been applied to idealized atmospheric problems; yet forecasting had become increasingly formalistic, divorced from any physical understanding of the processes responsible for weather change. So called empirical methods had mushroomed during the late nineteenth century; forecasters, by correlating weather phenomena, arrived at these methods through experience. By 1900 most meteorologists sought statistical patterns rather than physical or dynamic insights to predict weather. Disillusionment set in. Institutional conservatism toward new approaches and ideas tended to reinforce a sense of hopelessness. One observer complained, "Most meteorologists appear to be like the old Jesuit who refused to look in Galileo's telescope because he was afraid to see spots on the sun. . . . The foreign meteorological institutes stubbornly hang on to their old useless [*odugliga*] method." Some meteorologists even abandoned the belief in the possibility of scientific weather prediction.[6]

Two events, however, brought new hope, at least to those meteorologists not paralyzed by disillusionment: the advent of powered flight and World War I. In the prewar years a subdiscipline developed within meteorology which focused on the air aloft, accessible with balloons and kites. So-called aerologists, who often came from outside meteorology proper, benefited from the growth of aeronautics; they aimed at using aeronautical devices for the "scientific study of the free air." Soon the upper air no longer interested just aerologists; most meteorologists had to consider weather conditions aloft once the war began. The need to know when to launch a gas attack, whether planes could safely fly to and bomb an enemy position, or how to aim long-range artillery units changed the notion of forecasting. Field weather services used wireless telegraphy and mobile telephone systems to institute frequent and rapid exchanges of detailed weather information.

After the war, commercial aviation required similarly rapid but even more detailed and precise exchanges of data and forecasts. Meteorologists began to consider previously ignored weather phenomena and to introduce into their observations an unprecedented

6. Nils Ekholm describes his frustrations to Svante Arrhenius, 12 January 1906, SAP; Bjerknes relates Henrik Mohn's progression from optimism to disillusionment to Hugo Hergesell, 30 June 1926, copy, VBP and in his "Mindetale over professor dr. H. Mohn," *Forhandlinger i Videnskapsselskapet i Christiania Aar 1916* (Christiania, 1917), p. 114; William Napier Shaw, "A Century of Meteorology," *Nature* 128 (1931), 925–26.

refinement in classification of cloud forms and weather types. Increasingly, meteorologists, rather than simply issue warnings of dangerous weather conditions, emphasized a resource perspective toward weather: coordinating commercial and military activities, and especially flight with weather conditions could increase efficiency. To various degrees government agencies and meteorologists shared this perspective of weather, which promised substantial disciplinary growth. But meteorologists wondered to what extent they could and should modify their forecasting practices, their dependence on theory, their conceptual foundation to meet the tasks at hand. For those meteorologists and institutions mired in years of rote forecasting, the response was only an expansion of earlier formalistic methods. Others attempted completely new strategies.

Vilhelm Bjerknes well understood the changing conditions for atmospheric science and weather forecasting, capitalized on them to develop new methods for studying and using weather, and campaigned to disseminate these methods around the world. Bjerknes acted. He assessed situations. He devised, implemented, and when expedient, modified research programs and professional strategies. True, during the period 1900–1925 enormous changes arose in the social significance of weather. But to understand the transformation of meteorology we must know why Bjerknes recognized changing conditions for advancing atmospheric science and how he responded to them. I hope to show not an impersonal process—capital pressing science into its service—but a more complex relation among science, society, and the individual researcher.

Bjerknes was not of course free to choose his actions if he desired professional success. His goals and research programs, like those of other scientists, were subjected to market forces both in professional disciplines and in society. He recognized the importance of deliberating on what problems, analytic techniques, and research programs could command attention, attract resources, and bring professional rewards. Bjerknes's need for professional success indeed underlay the transformation of meteorology, but the conditions for pursuing his goals both directed and informed the result.

Bjerknes has been described with great affection by his former assistants, and rightly so.[7] His charm and skill as a lecturer, his informal leadership, sense of humor, unfailing devotion to his assistants,

7. Tor Bergeron, Olaf Devik, and Carl Godske provide the standard literature; see their accounts, edited by Arnt Eliassen and Einar Høiland, "Vilhelm Bjerknes: March 14, 1862–April 9, 1951," *GP* 24 (1962): *In Memory of Vilhelm Bjerknes on the 100th Anniversary of His Birth*, 7–37. A list of Bjerknes's publications can be found on pp. 26–37.

and his ceaseless energy and determination are the personal qualities often underscored in hagiographic literature. But Bjerknes, like any person, was complicated. His letters to family and closest scientific colleagues reveal a more nuanced—and realistic—picture. Naturally, in a life marked by disappointments, expressions of frustration, self-doubt, peevishness, and at times, bitterness do appear, but his ability to draw repeatedly upon inner strength and willpower to persevere in spite of obstacles is not easily explained.

His self-image and his historical articles are not necessarily reliable as primary factual sources. He wrote much of this literature for debates and discussions on science policy in Norway and on priority disputes. These articles are very skillfully crafted documents, constructed to persuade. Bjerknes was what we today might call a disciplinary entrepreneur, in the finest sense of the term. He ventured to build and develop new institutions and new science; he struggled to improve the conditions for research in Norway; he endeavored through science to improve society. His desire as a young scientist to obtain international recognition by formulating research programs of visionary scope and extraordinary difficulty may well have been unrealistic. But without such fantasies, the history of science would be quite different indeed.

To explore this history I first examine in Part I the circumstances leading to Bjerknes's formulation in 1903 of a research program for creating an exact physics of the atmosphere. Bjerknes did not gladly choose to become "the father of modern meteorology," as he has come to be known. To understand why Bjerknes, a physicist, eventually devoted his career to this pursuit and to understand his vision of a new meteorology, I discuss his entry into the world of professional science. Attitudes and lessons he learned in his early career he rarely forgot when he planned strategies later in life. In Chapter 1, I offer an account of the rise and fall of Bjerknes's professional expectations in physics, from when he began as an assistant to Heinrich Hertz in 1890 to when he decided to abandon theoretical physics in 1906. I detail Bjerknes's frustration at pursuing a scientific career on the geographic-cultural periphery, especially at his inability to command resources to develop an influential school of thought. In Chapter 2, I trace Bjerknes's initial contact with geophysical research and reluctance to make a professional commitment to it. Bjerknes eventually devised and devoted himself to a plan for creating an atmospheric physics, in part to defend the legitimacy of a classical mechanical world view for physics, in part to try a new tack for making a name for himself in science. He aimed first and foremost to invent graphic methods with which to apply hydrodynamic and thermodynamic principles to the atmosphere; he sought "rational predictions" of atmospheric changes. His plan to effect a conquest of the air by science was designed to coincide with the coming conquest of the air by aeronautics.

In Part II, I focus on Bjerknes's attempts from 1906 to 1917 to pursue this research program. To attain the data and the assistance to begin this task, Bjerknes involved himself and the project with the growing aerological specialty that explored the upper atmosphere with balloons and kites (Chap. 3). He campaigned to introduce absolute cgs units of measurement into aerology, and meteorology in general, on the grounds that his project's methods could be of assistance to aeronautics. By moving in 1907 from Stockholm to Christiania (not called Oslo until 1924) and then in 1913 to Leipzig, Bjerknes learned that aeronautical concerns held the greatest promise for obtaining the resources needed for his project (Chap. 4). The remarkable initial achievements of his budding Leipzig school were cut short by the war.

Bjerknes made several adjustments to the war. He moved to Bergen, on Norway's west coast, and began adapting his project to conditions then prevailing in Norway and to those he expected to come about there and in Europe after the war. I show how Bjerknes transformed his goals and project: first tentatively in reaction to wartime exigencies in neutral Norway, especially the threat of famine (Part III), and then more decisively in reaction to expectations of postwar commercial aviation (Part IV, Chap. 7–8). He turned to practical weather forecasting. New innovative forms of forecasting practice prompted and made possible the development in 1918 and 1919 of a preliminary new cyclone model based on three-dimensional surfaces of discontinuity separating differing air currents. Starting with a discovery by Bjerknes's son Jacob during forecasting, the Bergen scientists shaped and modified this model over a several-month period. I turn in Chapter 8's conclusion to the question of the historical specificity of the 1919 Bergen cyclone model, by which I mean the historical conditions that made it possible to establish the model; these first arose after 1914.

Experience had taught Bjerknes to take active measures to spread his methods and ideas. Amid postwar enthusiasm for meteorology's rapid growth as a discipline, he launched campaigns to attract disciples and to spread the preliminary methods and cyclone model. In Chapter 9, I show how these efforts prompted the Bjerknes group during the 1919–20 winter to constitute the polar front concept in reaction to aviation's meteorological needs. Polar front meteorology, which encompassed the group's earlier achievements, gave the young Bergen school an identity through which it sought international leadership. Spreading the Bergen meteorology abroad required keeping conceptual details simplified, and this necessity led members of the group to differ on the reality of several subsequent findings and on the need to modify their initial models.

Surviving at home made opposite demands on the Bergen school. Severe economic depression in Norway beginning in 1920 threatened the entire enterprise. Opportunities arose, however, for maintaining the new weather service for West Norway. In Chapter 10, I show that to integrate weather into commerce and, above all, fishery, Bergen meteorologists continued to sharpen and improve their forecasting practices; one result was the so-called occlusion process and in turn an evolutionary cycle for cyclones.

In the final chapter, I reach back to the summer of 1918 to follow the development of the Bergen air-mass concept in response to the need to forecast weather phenomena not associated with cyclones or fronts. Specific exigencies in Norwegian agriculture made the problem of forecasting local afternoon showers especially acute at this time; moreover, for aviation the need to forecast these and other "fair-weather" problems such as turbulence, gustiness, and visibility became increasingly apparent after 1920. By 1925 the Bergen school had established a method for classifying air masses on the basis of their life histories. This accomplishment, with the other achievements, enabled its members to predict with greater certainty and precision weather phenomena critical to several commercial spheres. Also by this time, Bjerknes and some of his assistants had begun the long, arduous task of placing the Bergen school's preliminary findings on a firm theoretical foundation.

PART I

APPROPRIATING THE
WEATHER FOR PHYSICS:
A PROFESSIONAL STRATEGY
(1892–1906)

[1]

The Rise and Fall of
a Career in Physics

A Mechanical World View for Physics

Vᴵᴸʜᴇʟᴹ's father, Carl Anton Bjerknes (1825–1903), professor of applied and later of pure mathematics at the Royal Frederik University in Christiana, had long been fascinated with the idea of showing that action-at-a-distance forces were actually forces produced and propagated in a fluid-medium–filled space.[1] Intrigued since his youth by Leonhard Euler's *Letters to a German Princess,* in which the great eighteenth-century natural philosopher and mathematician argued against Newtonian forces acting in empty space, the elder Bjerknes first began to work on this problem after hearing the Göttingen lectures of P. G. Lejeune Dirichlet in 1855. While developing new and powerful forms of mathematical analysis, Dirichlet considered a host of idealized physical problems, including that of bodies moving in fluids. The elder Bjerknes learned that a sphere moving with uniform motion in an ideal, frictionless, incompressible fluid experiences no resistance, just as if it were moving in empty space. He could then reject the argument against a medium-filled space—that it would hinder motion—and return to the thoughts Euler had inspired. By considering whether the principle of inertia in dynamics could be maintained in a system of bodies in a fluid medium, he gradually formulated a series of problems that led him to analyze

---

1. Vilhelm Bjerknes, "Til minde om professor Carl Anton Bjerknes," *Forhandlinger i Videnskabsselskabet i Christiania Aar 1903,* no. 7 (Christiania, 1904), 1–24; idem, *C. A. Bjerknes: Hans liv og arbeide: Træk av norsk kulturhistorie i det nittende aarhundre* (Oslo, 1925), pp. 37–44, 82–102, 143–64; *Studentene fra 1880* (Christiania, 1905), pp. 47–48.

hydrodynamic analogies to various action-at-a-distance forces: Would the motions of two bodies in a fluid be independent of each other? Would the motion of one produce in the other a motion that resembled electric, magnetic, or gravitational action at a distance? Thus, he began a lifelong attempt to describe apparent action-at-a-distance forces as forces arising within and propagated through a fluid medium.

But Carl Anton Bjerknes worked in isolation. In the context of Continental physics during the 1870s and 1880s his attempts to eliminate action-at-a-distance forces were at odds with such triumphs as those of F. E. Neumann and Wilhelm Weber, who established traditions for mathematical physics based on Newtonian-type forces. Nevertheless, when at the 1881 Paris International Electric Exhibition, he—and Vilhelm—demonstrated instruments that reproduced hydrodynamic analogies, few observers could ignore these baffling phenomena. Such celebrities as Hermann von Helmholtz, Gustav Kirchhoff, William Thomson (Lord Kelvin), the Siemens brothers, and the Marquis of Salisbury visited the small Norwegian exhibit booth and watched with amazement as a system of pulsating spheres and similar devices appeared to reproduce well-known electric and magnetic phenomena.[2] For many observers the Bjerknes apparatus seemed to illustrate that the mysterious nature of electricity could perhaps be revealed. British observers allegedly exclaimed, "Maxwell should have seen this!"[3] Of the eleven *diplômes d'honneur,* seven went to non-French exhibitors, including Werner Siemens, Thomas Edison, Alexander Graham Bell, and William Thomson. Professor Carl Anton Bjerknes, representing Norway, joined their ranks.

After his success in Paris, Carl Anton began receiving requests for instruments and for demonstrations from the international physics community. Further hope that his work could achieve recognition arose in the wake of Heinrich Hertz's extraordinary experiments. Drawing on a series of experiments conducted between 1886 and 1888, Hertz claimed to have shown the existence of electromagnetic waves, finitely propagated through space, as predicted by James Clerk Maxwell. Fame and accolades for Hertz soon followed. Many Continental physicists now acknowledged the need to abandon action-at-a-distance forces for comprehending electromagnetic phenomena and to develop a depiction based on contiguous actions in the ether. Bjerknes's studies could in principle play an important role. But perfectionism and self-criticism, lack of training in recent elec-

2. Vilhelm Bjerknes, *C. A. Bjerknes,* pp. 165–89.
3. Ibid., p. 179.

tromagnetic theory, and recurring writing blocks prevented the elder Bjerknes from preparing a manuscript. Carl Anton needed assistance to complete the work on analogies and to write a book; he recognized his efforts would otherwise remain incomplete and unable to exert influence.[4]

Although Vilhelm had to think of his own career, from the start he accepted responsibility for helping with his father's research program. As a youth he watched and helped his father devise simple experiments for displaying and studying the hydrodynamic analogies. As a student he designed and built instruments for his father; these he displayed with great pride at several universities and at the 1881 Paris exhibition, where he, not his father, explained and demonstrated the analogies to the delight of the spectators. Finally, he was drawn toward studies that could directly or indirectly assist his father's efforts. True, he had to get away from his father at times to develop his own skills and career opportunities: cramming in isolation for his university examinations and applying for a fellowship to study abroad during the 1889–91 academic years. Still, his father's work was never far from his thoughts.[5]

During his two years abroad Bjerknes began to learn through direct experience how the world of professional science functioned. In Paris he followed the lectures in mathematics (Charles Hermite and C. E. Picard), mechanics (P. E. Appel and Henri Poincaré), and mathematical physics (E. E. Mascart and Poincaré). He also followed with special concern a growing controversy on the validity of Hertz's investigations. In 1889 some doubts arose in France.[6] After the discovery by the Genevan physicists Lucien de la Rive and Edouard Sarasin of "multiple resonance phenomena," the French physicist M. A. Cornu used these results to argue for a rejection of Hertz's findings on the ground that the primary conductor did not have a fixed period of oscillation. In his lectures on electromagnetism at this time, Poincaré discussed Cornu's attack and even expressed his own reservations about Hertz's results.[7] The French attitudes shocked Bjerknes. He wrote to his father, "Here in France it is now fashionable to disparage Hertz." It seemed to Bjerknes that many French scientists took special

4. Vilhelm Bjerknes, "Videnskabelig selvbekjendelse" (hereafter cited as "Vid. selvbekj."), October 1910, posthumous papers 9a, VBP.

5. Ibid.; *Studentene fra 1880*, pp. 47–48.

6. Bjerknes's letters to his father (BFC), 1889–90, provide details of the Parisian physics and mathematics communities and of French reactions to Hertz.

7. These reservations were largely removed when the lectures were published; discussed in V. Bjerknes, "Vid. selvbekj."; Bjerknes's lecture notes, "Poincaré: Forelæsninger om elektrodynamik andet semester 1890," Dept. of Physics Library, cat. no. G:733, UBO.

pleasure in criticizing the young German celebrity. One "supposedly intelligent young physicist" claimed that anyone who had read Maxwell could have done the same; others simply asserted that Hertz had not taught them anything new or "proudly cited" Cornu's remarks. Bjerknes also noted that when he ordered an entire year's edition of *Wiedemann's Annalen* from the university library, the entire volume still had uncut pages.[8] While following Poincaré's lectures and the debate on the validity of Hertz's conclusions, Bjerknes wrote to Hertz asking whether he could study in his laboratory at Bonn University. In addition to obtaining research experience in a physics laboratory, which had been impossible at the relatively impoverished university in Norway, he especially wanted to work with Hertz, whose experiments he thought to be "the most beautiful" devised in the history of physics. Hertz invited Bjerknes to come to Bonn.[9]

Bjerknes saw the importance for his father's work of trying to defend Hertz's claims concerning the existence of finitely propagated electromagnetic waves in an ether medium. When he arrived in Bonn, Bjerknes received several surprises. To his shock and despair, he was the only student registered to work in Hertz's laboratory, which was poorly equipped and had suffered years of neglect. Having only recently come to Bonn, Hertz was just beginning to fix it up.[10] On the positive side, Hertz suggested research for Bjerknes that could assist in answering the French criticism. In Paris he had learned from Sarasin that Hertz could not defend his claims with the aid of new experiments because he had damaged his eyes while conducting the original painstaking investigations.[11] Bjerknes, instead of spending his evenings at a physics *kneip*, drinking beer and discussing science, as he had imagined, found himself, with the lack of colleagues and the challenge entrusted to him, spending long hours alone as he built equipment to study resonance phenomena for Hertz.

After initial difficulties and fears that Hertz was losing patience, Bjerknes managed to explain both experimentally and theoretically that the multiple resonance phenomenon was actually the consequence of strong damping of the oscillations in the primary conductor. Apparently Hertz had intuited just such a result, but without precise measurements he withheld public comment.[12] Just when Bjerknes achieved this result and was ready to rejoice, he learned

8. V. Bjerknes to C. A. Bjerknes, 20 April 1890, BFC.

9. Ibid.; V. Bjerknes to Heinrich Hertz, 20 December 1889, DM; Hertz to V. Bjerknes, 31 December 1889, copy, VBP (Bjerknes returned Hertz's letters to the Hertz family; the originals are in DM).

10. Bjerknes to C. A. Bjerknes, 20 April, 11 July, 2 November 1890, BFC.

11. Ibid., 20 April 1890.

12. Ibid., 4, 16 February, 29 April 1891.

from Hertz that Poincaré had come to the same conclusion and was ready to publish his findings. Bjerknes was crushed. His anguish was the greater because a Swedish colleague in Paris had informed Poincaré about Bjerknes's investigation but waited several weeks before letting Bjerknes know about the Frenchman's eager interest in and activity on the same problem.[13] Bjerknes had been working slowly and methodically; finally, under great stress, he finished a manuscript and with Hertz's help had it sent for quick acceptance in *Wiedemann's Annalen*. Poincaré's manuscript had been accepted several weeks earlier for publication by *Archives des Sciences Physiques et Naturelles de Génève*.

Although Poincaré had demonstrated his proof only mathematically, Bjerknes felt that his own experimental-mathematical investigation—and the past several months of anxiety—had been for naught: Poincaré, the well-established researcher, would get the credit for discovery and for putting an end to the conflict over Hertz's claims. Bjerknes pressed on nonetheless with his research during the spring and determined that the oscillations in the primary conductor are damped in a sine curve. He won the respect of Hertz, who hoped that Bjerknes would take a German doctorate and apply for a position at a German university, where he could continue these investigations. Instead Bjerknes chose to return to Norway, where there were few chances for a research career. His father had urged him to come home.[14]

Although Vilhelm Bjerknes's professional plans were uncertain, his scientific outlook was soon fixed. In Norway he managed under primitive conditions to continue the line of research he had begun in Bonn: to use resonance phenomena quantitatively to determine physical constants for the primary and secondary Hertzian conductors. Toward this goal he studied the effects of metals on the propagation of electric waves. He obtained a Norwegian doctorate in 1892, based on his investigations of Hertzian waves.[15] As part of his doctoral disputation he presented a public lecture, "On the Application of the Principles of Mechanics in Physics," in which he conveyed the optimism with which physicists—and he—regarded their science.[16] During the past century, he pointed out, several previously separate divisions within physics had been united: radiant heat studies had

13. Ibid., 4 June 1891.
14. Hertz to V. Bjerknes, 11 November 1891, copy, VBP; C. A. Bjerknes to Gösta Mittag-Leffler, 29 March 1892, GML; V. Bjerknes, "Vid. selvbetj."
15. Vilhelm Bjerknes, "Om elektricitetsbevægelsen i Hertz's primære leder," *AMN* 15 (1892), 165–236.
16. Vilhelm Bjerknes, "Om anvendelsen af mekanikens principer i fysik," ibid., 331–47.

eliminated the conceptual boundary separating heat and light; electromagnetism combined electricity and magnetism while also bringing optics into this domain. The advantage of such linkage was that obscure phenomena could be associated with previously well-understood phenomena, and linking different divisions of physics to mechanics had proved the most auspicious. Bjerknes foresaw that this process of eliminating conceptual boundaries between the branches of physics would continue until all these divisions "have fused together into a single entity, and this entity is mechanics."[17] In the meantime, Bjerknes's studies on electromagnetic waves began attracting considerable attention abroad. Carl Anton, however, was urging his son to devote more time to hydrodynamic analogies, Vilhelm understood that the prospects of obtaining a permanent position in Norway which would permit research were dim at best.

Carl Anton realized his only chance of seeing his many years of work published would be for Vilhelm to assume responsibility for preparing a manuscript.[18] Fearing that Vilhelm might return to Germany and to full-time investigations of electric waves, Carl Anton turned to his colleague at the Stockholm Högskola, the mathematician Gösta Mittag-Leffler. Although Mittag-Leffler would have preferred someone with greater mathematical skills for the new lectureship in mechanics, he nevertheless helped Bjerknes obtain this position, which offered considerable opportunity for research.[19] The Stockholm Högskola had been founded in 1878 as an alternative to the state-run older universities in Lund and Uppsala, which seemed to many to be hopelessly bogged down in regulations and conservative traditions. The founders of the högskola, many from the growing industrial bourgeoisie, looked to the Collège de France and the Royal Institution of London as models for the new institution: pursuit of knowledge, rather than preparation for civil service examinations, was to be the ideal. Private donations kept the högskola functioning but also meant slow and sporadic growth.[20]

Bjerknes arrived in Stockholm early in 1893 with folders of notes, data, and drafts of unfinished articles—all belonging to his father. To

17. Ibid., p. 347.
18. C. A. Bjerknes to Mittag-Leffler, 29 March, 17 October, 19 December 1892, GML.
19. Ibid., 17 October, 19 December 1892, 6 October 1893; V. Bjerknes to Mittag-Leffler, 14 October 1892, GML; V. Bjerknes to "Svigerfader" [father-in-law, Jakob Aall Bonnevie], 8 May 1893, BFC.
20. *Stockholms högskola 1878–1898: Berättelse öfver högskolans utveckling under hennes första tjugoårsperiod, på uppdrag of hennes lärareråd utgifven af högskolans rektor* (Stockholm, 1899), esp. pp. 1–34; Fredrik Bedoire and Per Thullberg, *Stockholms universitets historia 1878–1978* (Stockholm, 1978), pp. 16–36; Sven Tunberg, *Stockholms högskolas historia före 1950* (Stockholm, 1957), pp. 2–38.

Hertz, he wrote that he would now lecture on hydrodynamic action at a distance as a means of beginning to systematize and complete his father's studies, and although he would work whenever possible with theoretical studies on electric waves, further experimental studies would have to wait until he obtained competent assistants. Being rather optimistic, Bjerknes thought that by using his lectures to organize his father's materials, he might be ready to publish in about a year.[21] He was not planning to turn his back on the research opportunities arising from his work with Hertz. In fact Bjerknes still managed to continue publishing on electric waves and soon also began supervising a student, Nils Strindberg, with experiments on resonance phenomena. Having finally found an elegant quantitative method for using resonance phenomena to determine the fundamental constants of Hertzian conductors, Bjerknes considered applying this method to the study of energy radiation. Hertz expressed great satisfaction with Bjerknes's investigations; international attention came readily.[22] But Carl Anton expressed concern that these studies distracted Vilhelm from completing *their* manuscript.[23] Finally in 1895 after struggling to be named to a new professorship in mechanics and mathematical physics, Bjerknes abandoned—with anguish—all research related to electric waves and dedicated his efforts exclusively to hydrodynamic actions at a distance.[24]

In addition to filial loyalty, this refocusing of interest and goals can be attributed in part to the scarcity in Stockholm of laboratory facilities and students with which a major experimental research project could be pursued and in part to Hertz, who had recently died. Bjerknes, rather than merely compiling and editing his father's investigations, considered expanding this work into a comprehensive mechanical physics based on contiguous actions. In his posthumously published *The Principles of Mechanics, Presented in a New Form* (1894), Hertz had proclaimed the need to recast mechanics in a form compatible with the recent advances in physics.[25] Because instantaneous action-at-a-distance forces could serve no longer as a basis for com-

21. V. Bjerknes to Hertz, 11 April, 10 May 1893, DM.

22. V. Bjerknes, "Vid. selvbekj."; Hertz to V. Bjerknes, 18 April, 6 June 1893, copies, VBP.

23. C. A. Bjerknes to Mittag-Leffler, 15 July 1894, GML.

24. V. Bjerknes, "Vid. selvbekj."; V. Bjerknes to Philipp Lenard, 11 May 1903, 5 April 1904, PLP; on opposition to Bjerknes within the högskola, see Svante Arrhenius to Edward Hjelt, 26 July 1895, EHP, and documents related to the naming of new professors in physics and in mathematical physics (for which Arrhenius and Bjerknes respectively applied) in the Stockholm Högskola archive, Riksarkivet, Stockholm; the small faculty was divided into factions; its disputes at the time, particularly furious.

25. Russell McCormmach, "Heinrich Hertz," *DSB* 6:340–50, places Hertz's investigations in historical context; see also Martin J. Klein, "Mechanical Explanation at the End of the Nineteenth Century," *Centaurus* 17 (1972), 58–82.

prehending electromagnetic phenomena, he hoped to recast mechanics in a form based on contiguous actions in an ether. To show the feasibility of such a mechanical-ether physics, he derived the principles of mechanics using only the concepts of mass, space, and time. Rather than employing the notion of force or energy, he postulated that in addition to visible masses there exist moving, hidden masses that are bound to one another by rigid constraints. With this supplementary hypothesis he derived axiomatically the lawful behavior of perceptible mass without introducing force or energy transformations. True, Hertz's contemporaries generally concurred with the book's opening declaration, "All physicists agree that the problem of physics consists in tracing the phenomena of nature back to the simple laws of mechanics." True, many physicists regarded Hertz's book as elegant and beautiful. But they quickly discovered its limited value, for Hertz did not provide physical examples of his abstract mechanics. Bjerknes, however, recognized that for his and his father's research program the book had great significance; he was overjoyed when he received a copy.[26]

Hertz's aim of finding a mechanical basis for a contiguous-action depiction of electromagnetism coincided with Bjerknes's research interest. Hertz's call for establishing a new representation for mechanics, which could serve to establish a physics of the ether, suggested a broad, comprehensive goal toward which Bjerknes could strive. By providing a physical example of the mechanics that Hertz derived only axiomatically, Bjerknes believed he could help lead the way to a new epoch in mechanics and mathematical physics.[27] In his view, hydrodynamic actions at a distance were an illustration of Hertz's mechanics; his father had been working toward the same goal as Hertz, only beginning with specific phenomena rather than postulates and axioms. The pulsating spheres and rotating cylinders that exhibit attractions and repulsions would appear to be acting on each other as if by instantaneous action-at-a-distance forces across empty space if the fluid in which they interacted was not readily visible. Consequently, Bjerknes saw the opportunity to transform his and his father's hydrodynamic investigations from a popular curiosity, which few scientists adopted as a problem for inquiry, into a cornerstone for a new mechanics and mathematical physics. And this endeavor would carry the authority and prestige of Hertz's name.

26. V. Bjerknes to C. A. Bjerknes, 10 September 1894, BFC.

27. V. Bjerknes, *Vorlesungen über hydrodynamische Fernkräfte nach C. A. Bjerknes' Theorie* (Leipzig, 1900), 1:6–8, 272–74; idem, "Til minde om prof. C. A. Bjerknes," pp. 11–12; "Les actions hydrodynamiques à distance d'après la théorie de C. A. Bjerknes," *Rapports présentés au Congrès international de physique réuni à Paris en 1900* (Paris, 1900), pp. 253, 273–74; V. Bjerknes to C. A. Bjerknes, 28 May 1899, BFC.

A Generalized Circulation Theorem

When in 1895, Bjerknes withdrew from active work on electric waves to prepare a systematic presentation of his father's investigations, he began his new lecture series on hydrodynamic action at a distance with a two-semester analysis of Hertz's *Mechanics*.[28] To make the hydrodynamic analogies to electric and magnetic phenomena more convincing, he tried to derive the relationships independently of the geometric shape of the pulsating or rotating bodies.[29] This particular problem vexed Bjerknes for several years and eventually led him to formulate generalized theorems for circulation and vortex motions, which in turn proved decisive for his future.

To generalize the hydrodynamic action-at-a-distance analogies, Bjerknes postulated a situation in which the pulsating and oscillating bodies are part of the fluid medium; that is, he imagined fluid bodies differing in density and compressibility from the surrounding fluid. He hoped to derive equations that would be independent of the specific shape of the bodies and to consider situations analogous to various electromagnetic phenomena. Difficulties ensued. Attractive and repulsive phenomena between the fluid bodies seemed to occur with an accompanying production of vortices in the boundary layer between the fluid bodies and the surrounding fluid.[30] This result contradicted the well-established theorems of Helmholtz and Lord Kelvin which claimed vortex motions and circulations in frictionless, incompressible fluids are conserved: In their view, such motions, left to themselves, neither die away nor arise in these idealized fluids. Kelvin even claimed the validity of his theorem for compressible fluids. Yet, Bjerknes's mathematical results suggested the production of such motions.

Persuaded that a complete set of dynamic analogies between hydrodynamic and electromagnetic forces must exist, Bjerknes refused to abandon his results. He could find in *The Principles of Mechanics* further justification for refusing to accept the earlier theorems as definitive, for there Hertz questioned the need to accept Kelvin's simplifying assumptions about the nature of a fluid if one is to derive a

28. "Hertz' mekanik" and "Hertz' mekanik fortstat," lecture notes for fall 1895 and spring 1896 semesters, black folder 1, 2, VBP. Throughout his life Bjerknes considered Hertz's *Mechanics* the starting point for modern physics: V. Bjerknes, "Om veirforutsigelse som fysisk problem," *Naturen* 47 (1923), 33; idem, "Forelæsning 9 April 1907, Tiltrædelsesforelesn.," posthumous papers 9a, VBP.

29. V. Bjerknes, "Vid. selvbekj."; idem, *Vorlesungen über hyd. Fernkräfte* (Leipzig, 1902), 2:175; V. Bjerknes to C. A. Bjerknes, 18 December 1895, VBP.

30. V. Bjerknes, "Vid. selvbekj."; idem, "Fra de pulserende kuler til polarfronten," *Ymer* 59 (1939), 190–91.

mechanics of the ether.[31] So Bjerknes ventured to find a flaw in the classical hydrodynamic studies of Helmholtz and Kelvin, which also included investigations of analogies between hydrodynamics and electric or magnetic phenomena.[32] Realizing that Helmholtz and Kelvin had earlier considered only geometric analogies arising from the vector nature of both sets of phenomena, Bjerknes claimed they had missed the possibility of deeper, more significant, dynamic analogies, so their results should not be taken as definitive. He continued to analyze two specific issues with greater resolve. On the one hand he aimed to transcend geometric figures by starting from an analysis of two cylinders rotating in a fluid as a situation similar to two electric circuits in a parallel field. On the other he continued investigating mathematically instances of fluid bodies pulsating or vibrating in a fluid of differing density. This he did by considering physical models of how the individual heavier and lighter particles of the fluids behave to produce attractive and repulsive phenomena.

Finally, walking home from the högskola late in March 1897, Bjerknes suddenly realized that Kelvin's extension of conservation of circulation to compressible fluids did not apply to his own problem. In Kelvin's instance the density of the fluid depended only on pressure. But Bjerknes postulated a heterogeneous fluid without restrictions on compressibility so that, in principle, density could be a function of several variables and not simply of pressure alone.[33] His own model was based on new and unexplored phenomena: circulation could apparently arise from the fact that density is a function of several variables in addition to pressure. Once Bjerknes overcame this obstacle, he proceeded very quickly to derive equations similar to those of Helmholtz and Kelvin except that the change of rate of circulation, or of rotation, with time was not zero. Although the mathematical steps proved quite simple, Bjerknes regarded the implied physical theory as a breakthrough.

The existence of dynamic analogies between such dissimilar phenomena as hydrodynamics and electromagnetism implied a deeper, common origin, which Bjerknes had always assumed.[34] His firm belief that hydrodynamic analogies must exist for all electromagnetic

31. Heinrich Hertz, *The Principles of Mechanics, Presented in a New Form*, trans. D. E. Jones and J. T. Walley (1899; rpt. New York, 1956), pp. 37–38.

32. V. Bjerknes to C. A. Bjerknes, 21 February, 15 March 1897, BFC; V. Bjerknes, "Foredrag ved Höiskolens aarsfest 3/2 1897: Hydrodynamiske fjernkrefter," posthumous papers 9a, VBP.

33. V. Bjerknes, "Fra de pulserende kuler til polarfronten," pp. 190–91; reported by Bjerknes to the Stockholm Physics Society, "Fysiska sällskapet," *Vårt land*, 29 March 1897.

34. V. Bjerknes, "Vid. selvbekj."; V. Bjerknes to C. A. Bjerknes, 12 April 1897, BFC.

phenomena compelled him to go beyond the Helmholtz and Kelvin theorems and to derive generalized theorems for vortex and circulatory motions. After presenting these theorems in lectures on 9 and 10 April, he wrote enthusiastically about them to his father. He noted that he hoped also to find generalized equations for electric and magnetic phenomena so that he could ground both hydrodynamics and electromagnetism in a common mathematical foundation. His success gave him confidence "to write a book that all physicists will not only be able to read, but will have to read."[35] He began working strenuously and soon published two articles related to the circulation theorems while also preparing a manuscript for the book based on his father's investigations. In one of these articles Bjerknes focused on the potential application of his theorem to diverse atmospheric and oceanic motions, phenomena in which he had shown no previous interest. This sudden attention to geophysical questions even as Bjerknes was increasing his efforts both to complete a systematization of his father's work and to establish a new foundation for mechanical physics, is the subject of the next chapter. Before turning to Bjerknes's initial contact and gradual involvement with meteorology, an examination of his parallel endeavors in physics will assist in appreciating his career shift.

Professional Crisis

When Bjerknes resolved in the 1890s to broaden the mechanical foundations of physics, he believed that this work conceivably could establish him as a prominent physicist and vindicate his father's efforts spanning a quarter century. He was aware that he had turned away from a significant research field when he abandoned wave and radiation studies. By the end of the decade he saw his earlier results integrated into the growing research industry in this field as well as the growing industry related to wireless telegraphy. Although he was still being cited in the relevant literature, his professional reputation was becoming increasingly dependent upon the fate of his mathematical physics. He had taken a chance in changing the focus of his research; in some respects he had little choice, given his father's dependence on him for writing a manuscript. Striving to publish the book while his father was still alive and healthy enough to appreciate the effort, Bjerknes naturally also wanted to publish a book that would enhance his status as a physicist.

Bjerknes encountered problems in virtually all aspects of produc-

35. V. Bjerknes to C. A. Bjerknes, 28 May 1899, BFC.

ing his and his father's book. He conceived the work as a textbook that could be used as a basis for physics education: conceiving electromagnetic phenomena as dynamic analogues to hydrodynamic phenomena would allow the student to begin grasping the mechanical nature of electric and magnetic fields of force. Preliminary negotiations with the German publishing firm Johan Ambrosius Barth seemed auspicious. Interest in the book came quickly, especially after the recommendation and intervention of Philipp Lenard, who had been Hertz's primary assistant and who became Bjerknes's supporter and colleague. Both Lenard and Ludwig Boltzmann, to whom the publisher had turned for comment, recommended publication. In negotiating with the firm, Bjerknes did not want to appear too modest. He thought it best to negotiate from strength; as a tactic he stipulated royalties on sales. Arthur Meiner, the firm's agent, was a veteran of the trade. Using Boltzmann's characterization of the work as "difficult scientific" and pointing to the firm's losses on such texts as Helmholtz's *Vorelsungen*, Hertz's *Mechanik*, and Kirchhoff's *Abhandlungen*, Meiner not only brushed aside Bjerknes's request but proposed that the publishing ought to be subsidized. He promised, however, that the firm would offer extensive exposure for the book through its important journal *Annalen der Physik*.[36] Disheartened, Bjerknes did manage to negotiate an acceptable agreement for publication in two volumes.

Bjerknes, nervous and finding it hard to sleep, worked feverishly to complete the manuscript. He demanded leave of absence from teaching during the fall 1899 semester to be able to supervise the publishing of the first volume. Whether or not his request was justified (he wanted to be in Leipzig to check the equations, formulas, and translation into German), Bjerknes's closest colleagues at the högskola, Svante Arrhenius and Otto Pettersson, were dismayed and frightened by the tone and intensity of the request.[37] Referring to the "neuropathological" tone of the letter, Otto Pettersson feared Bjerknes was driving himself to the point of breakdown. He knew Bjerknes was driven to fulfill his duty as an obedient son, a duty in conflict with other obligations: "Now, it is best not to arouse conflicts between higher and lower duties within a person; such matters are best left to the writers of tragedy, Ibsen & Co."[38] Pettersson recommended extending the leave of absence, adding that of course should Bjerknes

36. Arthur Meiner (for J. A. Barth) to V. Bjerknes, 1 May 1899, VBP; several letters from Lenard to V. Bjerknes, May–June 1899, VBP; V. Bjerknes to C. A. Bjerknes, 18 May 1899, BFC.

37. Otto Pettersson to Svante Arrhenius, 18, 24 August 1899, SAP.

38. Ibid., 24 August 1899, SAP.

survive the publication, he would have a major scientific success on his hands. Bjerknes himself later admitted that he never again was able to work so intensely.[39]

While still preparing the manuscript, Bjerknes began to see threats to his plans. When he traveled to Leipzig in 1899 he heard firsthand about recent disputes and new discoveries in physics that he had been trying to follow from Stockholm.[40] As the foundation for comprehending physical reality, mechanics no longer claimed the undivided allegiance of physicists.[41] The capacity of mechanics to conceive of the ether and to explain electromagnetism was being called into question. So-called energeticists were asserting that energy transformations are the essence of physical phenomena and thus all that should be studied. Discussions with his supportive colleagues Max Planck, in Berlin, and Otto Wiener, in Leipzig, convinced Bjerknes to rewrite his book's introduction in a reconciliatory tone "to avoid all polemics."[42] Although he did not anticipate large-scale defections from the mechanistic world view of physics, he offered a cautious defense of his mechanical approach in light of the growing mood of skepticism. Taking a position in opposition to both the antimechanist extremes of energeticism and the excesses of those who "fantasize wildly" using mechanical models, Bjerknes advanced a comparative-phenomenological approach based on mathematical analogies: these could stand independent of depictions of the ultimate nature of electricity and magnetism.[43] In spite of his uneasiness, Bjerknes allowed himself to hope for success.

During the next few years Bjerknes followed with alarm the growing

39. V. Bjerknes to Lenard, 5 April 1904, PLP; V. Bjerknes to C. W. Oseen, 22 November 1925, CWO.

40. Bjerknes's letters to his wife, Honoria, and to his father during the summer and fall 1899, BFC, provide accounts of his meetings with German scientists.

41. Tetu Hirosige, "Electrodynamics before the Theory of Relativity, 1890–1905," *Japanese Studies in the History of Science*, no. 5 (1966), 1–49; idem, "The Ether Problem, the Mechanistic World View, and the Origins of the Theory of Relativity," *Historical Studies in the Physical Sciences* 7 (1976), 3–81; Russell McCormmach, "H. A. Lorentz and the Electromagnetic View of Nature," *Isis* 61 (1970), 459–97; idem, *Night Thoughts of a Classical Physicist* (Cambridge, Mass., and London, 1982), pp. 16–35, 92–118, which depicts how physicists perceived their science at this time. My discussion of the state of physics circa 1900 is meant to be not a comprehensive or analytical account but a general background for understanding Bjerknes's dilemma.

42. V. Bjerknes to C. A. Bjerknes, 12 September 1899, BFC.

43. Ibid.; V. Bjerknes, *Vorlesungen über hyd. Fernkräfte* 1:1–8. Although Bjerknes tried to distance his work from "kinetic" mechanical models like Boltzmann's and from Arthur Korn's more extreme use of mechanical models, the Bjerkneses' hydrodynamic action-at-a-distance investigations were nevertheless often lumped with the work of those "impossible dreamers, who believe they have solved nature's riddles" (idem, "Vid. selvbekj.").

interest in nonmechanical foundations for physics. For some physicists the triumph for Maxwell's theory effected by Hertz's experiments suggested that the ultimate nature of physical reality is electromagnetic. Experimental results showing the existence of minute electric particles (electrons) reinforced nonmaterialistic views of physical nature. Some physicists claimed that if matter was in actuality electric, then mechanics ought be derived from electrodynamics; representations of the ether became increasingly dematerialized. In attempting to harmonize physicists' mathematical formalisms for continuous field representations, on the one hand, and particulate electric charges, on the other, the great Dutch physicist H. A. Lorentz elaborated an electron theory based on a perfectly stationary ether that represented a state of the electromagnetic field.[44] In contrast, the comprehension of electromagnetic phenomena with mechanical stresses propagated in a perfectly dragged ether, as ambiguously suggested by Maxwell, endorsed in some form by Hertz, and underlying Bjerknes's entire enterprise, seemed to be threatened.

By 1903, after his two volumes had been published (1900 and 1902), Bjerknes understood that this work was not going to have the impact that he had imagined. Most reviewers were polite about his two volumes; some were moderately enthusiastic, noting that Bjerknes had given the analogies the most complete treatment to date. Still others referred to the work as classic, in the sense of being complete and belonging to the past. Bjerknes was not pleased.[45] He conceded almost immediately that the book had come out at a bad time: just when attacks on the use of mechanics as a foundation for physics were becoming prevalent. For Bjerknes, as for so many of his colleagues, the early 1900s was a period of "scientific crises".[46] He maintained a conviction that Hertz's *Mechanics* eventually would assume the leading position in physics that it deserved, at which time his and his father's work would receive proper recognition.[47] He was not giving up hope just yet; he continued to promote a physics in which hydrodynamics, electromagnetism, and also elasticity might form a single entity: "I myself hold no doubt that this will become the future form for representation for mathematical physics, and quite independent of hypothetical explanations of electricity. . . . In this manner hydrodynamic

44. Russell McCormmach, "Hendrick Anton Lorentz," *DSB* 7:487–99.

45. Bjerknes first was excited by the attention the volumes received, perhaps not unlike any first-time author, but he soon recognized the nature of the praise and began to despair; V. Bjerknes to C. A. Bjerknes, 18 December 1902, BFC; V. Bjerknes to Lenard, 11 May 1903, PLP.

46. V. Bjerknes to Lenard, 11 May 1903, PLP.

47. Ibid.

phenomena will receive a central position in mathematical physics."[48] No one could then foresee physics' future development. Bjerknes, like many of his colleagues, continued to assume that the many scientific crises would be resolved within a mechanistic framework.

Bjerknes's hope that the pendulum of opinion would swing back toward mechanics was soon dashed. When Lorentz had come to Stockholm to receive his part of the 1902 Nobel Prize, he had mentioned Carl Anton Bjerknes's work in his Nobel lecture and informed Bjerknes that he was well acquainted with the first volume of the work on hydrodynamic analogies. Bjerknes then learned that Lorentz was preparing articles on Maxwell's theory for the *Encyklopädie der mathematischen Wissenschaften;* he was buoyed by the expectation that his and his father's efforts would there receive the attention and recognition hitherto unawarded.[49] Lorentz's contribution appeared in 1904. Perhaps Bjerknes misunderstood Lorentz's kind, patient, and undogmatic attitude as a critic. He did not know that Lorentz had received a strongly worded suggestion from Arnold Sommerfeld, the assistant editor, that at most the Bjerknes investigations, like other mechanical depictions of the ether, should receive not more than a passing comment.[50] Sommerfeld made it clear that the electromagnetic world view must receive prominence. The *Encyklopädie*, as it turned out, scarcely mentioned the Bjerkneses' hydrodynamic actions at a distance, which meant that the standard reference work on mathematical physics seemingly had condemned his and his father's efforts to obscurity.[51]

Even more disturbing for Bjerknes, Lorentz's most recent work on electron theory seemed to answer many of the criticisms leveled at the earlier versions of the theory. Together with publications by the more vociferous antimechanist adherents of an electromagnetic world view, such as Wilhelm Wien and Max Abraham, this development implied for Bjerknes that a mechanical world view could no longer be advocated in physics. Bjerknes understood full well that the question of world views was not simply a matter of philosophical preferences. His entire project rested on the notion of the ether as a mechanical entity in which electromagnetic phenomena are produced. He readily conceded that the analogies between hydrodynamic and electromagnetic

48. V. Bjerknes to C. A. Bjerknes, 23 October 1901, BFC; see also V. Bjerknes to Fridtjof Nansen, 18 January 1902, FNP; V. Bjerknes to J. A. Bonnevie, 20 September 1903, BFC.
49. V. Bjerknes to C. A. Bjerknes, 18 December 1902, BFC.
50. A. Sommerfeld to H. A. Lorentz, 21 March 1901, Lorentz Correspondence, film II, AIP. I thank Russell McCormmach for bringing this letter to my attention.
51. V. Bjerknes, "Vid. selbekj."

forces were far from perfected; he believed nevertheless that his father and he had established facts that physicists would have to include
in their corpus—that is, as long as physicists accepted the mechanical
presuppositions with which these facts could possess meaning. During
the summer of 1904, Bjerknes reviewed for himself the state of physics.[52] Finally, in desperation, he composed a letter to Lorentz, who as
the acknowledged leader of European theoretical physics and, as the
primary constructor of electron theory, could perhaps be persuaded
to acknowledge the continued tenability of a mechanical foundation
for physics.[53]

In his letter Bjerknes tried to reason that a perfectly dragged ether
and a mechanical world view could still be presupposed. He claimed
that he did not consider electron theory in itself problematic. His
complaint was directed to just one aspect of electron theory in its
present state: acceptance of a stationary ether threatened the entire
Maxwellian enterprise, for it excluded the possibility of mechanical
explanations of electromagnetic phenomena. Admittedly, the energeticists, who on principle did not want any mechanistic theories, perhaps
saw an advantage to such a strategy. Could not Oliver Heaviside's
attempts to continue the Maxwell tradition by exploring stresses in a
mechanical ether provide an alternative strategy to the dematerialized
ether? Bjerknes conceded that perhaps not all natural phenomena can
be explained mechanically. He accepted full well that an argument
could be made against mechanics as the ultimate fundamental reality
behind physical phenomena. But because the human senses and our
instruments allow us to comprehend these phenomena best mechanically, a mechanical world view could, he thought, facilitate comprehending physical reality. He opposed the tendencies to give up in
advance the *possibility* of constructing mechanical theories. According
to Bjerknes, the recent abandonment of the mechanical world view
should be seen not only as a problem of the "individual psychology of
physicists" but also as a problem of "mass psychology": in an ever-
growing popular movement, virtually everyone was rushing to join the
crowd in casting off the classical tradition of mechanics—at least so it
seemed to him. In a draft of this letter Bjerknes had claimed to being
"very conservative" with regard to these matters. For him, the stationary ether was nothing more than "a blackboard on which equations are
written"; he did not intend to relinquish his mechanical view of
nature.[54]

52. "Hydrodynamiske Analogier," black folder 15, VBP, contains Bjerknes's studies
of Lorentz's work written while on summer holiday in 1904.

53. "Udkast til et brev til Lorentz," black folder 15, VBP. The draft is dated August
1904; final version, 4 September 1904, HAL.

54. Bjerknes claims in a subsequent letter to Lorentz, 1 February 1906, HAL, to have
always considered the stationary ether "eine schwarze Tafel."

At the time, Bjerknes did not necessarily have to abandon his beliefs. Lorentz wrote back almost a year later and conceded that he himself accepted an absolutely stationary ether only as an approximation; his own electron theory would have to be altered significantly to adopt this postulate. He urged Bjerknes to continue with the hydrodynamic analogies and to retain a mechanical foundation for physics; he claimed that under certain conditions such a foundation might still be an alternative. Even if Lorentz was trying to be kind and diplomatic, he clearly did not believe in 1905 that any of the new approaches and syntheses were completely successful. He himself was still having difficulties in turning away from physics' nineteenth-century heritage. His comments brought Bjerknes "great pleasure".[55]

In fact Bjerknes was by no means alone in the world of physics. More than a few physicists were not quite ready to abandon a mechanical depiction of the ether as an explanatory basis for their science. German experimental physicists such as Lenard and Wiener, who were troubled by the increasing use of complex higher mathematics in electron theory, supported Bjerknes's efforts that relied on traditional mathematics in close contact with experimental demonstration. Others recognized in him a true disciple of Maxwell. Bjerknes was invited to inaugurate Columbia University's Adams Lectures, a series to be held annually with prominent European physicists discoursing on recent advances in theoretical physics.[56] American physicists responded positively to Bjerknes's lectures on the hydrodynamic analogies to electromagnetic forces. One enthusiastic supporter expressed a hope that they would learn as much from the next Adams Lecturer— Lorentz on the electron theory—and agreed with Bjerknes's comment, "There is more to do in physics than just electron theory."[57] The lack, among physicists, of consensus on their science's underlying foundations, as well as proliferating research topics (e.g., radioactivity and new forms of radiations) created possible niches for many networks of

55. Lorentz to V. Bjerknes, 11 November 1905, VBP; V. Bjerknes to H. Bjerknes, 27 November 1905, BFC. Quantum and relativity theories first attracted serious attention several years later.

56. Lenard to Nordahl Wille, 24 April 1906 and Otto Wiener to Wille, 27 April 1906, KUD/E, "Universitetet, Budsjetter 1905/06–07/08" (papers related to Bjerknes's appointment to the Royal Frederik University). Wiener's and Lenard's letters to Bjerknes contain frequent praise and encouragement. Oliver Heaviside's letters to Bjerknes show the Maxwellian fervor shared by both physicists. Bjerknes had been asked as early as 1902 by Michael Pupin to lecture on hydrodynamic action at a distance at Columbia University when the two met in Berlin (V. Bjerknes to H. Bjerknes, 11 July 1902, BFC).

57. A. P. Wills to Bjerknes, 13 March 1906, VBP. In letters to his wife, Honoria, December 1905, BFC, Bjerknes discusses American physicists' positive reception of his work: his lectures were interrupted by applause; he was elected a foreign member of the American Physical Society, joining Lord Kelvin and Svante Arrhenius.

investigators sharing research goals or orientations. Bjerknes could count on an audience and on support from some other physicists.

Bjerknes's problem as a physicist was not so much being alone in his view of physics as being alone in working on his specific research problem. His frequent complaint that he had published his two volumes at the worst possible time because "the energetic and electromagnetic world views" led physicists to regard mechanical theories with disfavor and distrust was only part of the problem.[58] Bjerknes also appreciated that he lacked professional authority and resources to persuade physicists to consider his problems and research program. Those colleagues who regarded his work favorably were generally either experimental physicists or investigators not then active in research. Leading European theoretical physicists did not take up his research program. What comfort was there in being called a leader in European physics if those offering praise did not themselves join his effort or encourage their students and colleagues to do so? How could he establish his own reputation, no less help his father's, if he could not convince other theoretical physicists of the value of his problems and methods of analysis, which were tolerated but certainly not prized?[59]

Professional isolation and disciplinary impotence, not rejection of his theories, plagued Bjerknes. He was, moreover, unable to change his research program or his basic orientation in physics. Throughout his life he felt compelled to justify his father's many years of hard labor, but even had he felt able to overcome the psychological-emotional links to this work, there remained overwhelming professional obstacles. He admitted in 1902 that he already had invested too much time and energy to abandon his search for a mechanical unification of physics.[60] By 1905, when he had acknowledged that the electromagnetic world view and electron theory were the leading edge of German-language theoretical physics, he surely also understood that the complex mathematics being integrated into physics to advance this work was beyond his technical competence.[61] Nor did Bjerknes find a return to experimental physics inviting. Since 1895, when he left the field, the developments had been many. Competition for prestige and success was keen. Bjerknes was not attracted to the kind of experi-

58. V. Bjerknes, "Vid. selbekj."

59. Pierre Duhem, Lenard, Lorentz, and Wiener referred to Bjerknes as either a leading or a first-class theoretical physicist in their respective letters on Bjerknes's behalf to Wille (see fn. 56).

60. V. Bjerknes to Nansen, 18 January 1902, FNP.

61. On the complexities of the mathematics, see Lewis Pyenson, "Physics in the Shadow of Mathematics: The Göttingen Electron-Theory Seminar of 1905," *Archive for History of Exact Sciences* 21 (1979), 55–89.

mental physics that dominated in Germany and Sweden: the "searching after the fifth decimal of one or another interesting constant according to laboratory pedantry's best precepts."[62] Regardless, the physics facilities at Stockholm Högskola were "miserable," and his growing professional and personal isolation in Sweden did not auger well for launching any major research program.[63]

Upon learning from Wiener that he might be called to a chair in mathematical physics at Leipzig University, he expressed his enthusiasm and underlying frustration in a letter to his wife Honoria: should he obtain this position, he then might be able to build a school of disciples and "force their theories upon the world!"[64] Bjerknes did not receive the post. At the högskola he had few chances to establish such a school, both because the institution did not receive a charter to grant degrees until 1904 and because there existed few positions in Sweden and Norway in which to place potential disciples. Moreover, in Sweden, Bjerknes felt "virtually alone" in physics.[65] His definition of mathematical physics, which might better be called theoretical physics, found few supporters in Sweden.[66] He called for avoiding sterile mathematical analysis and for seeking instead a deeper understanding of physical principles with the help of mathematics. In so doing, he lost any further support he could gain from Mittag-Leffler. In Uppsala and Lund, experimental physics largely defined the extent of the field. Bjerknes found that in all of Sweden he could find but one person with whom he could discuss matters related to electron theory.[67] Even publishing posed difficulties. To publish in the

62. V. Bjerknes to C. A. Bjerknes, 30 November 1890, BFC. Bjerknes expressed his dislike for classic German "laboratory pedantry" when relating how glad he was that he came to Hertz's laboratory rather than "the model laboratory in Strassburg."

63. V. Bjerknes to Lenard, 9 February 1906, PLP.

64. V. Bjerknes to H. Bjerknes, 11 July 1902, BFC; see also V. Bjerknes to C. A. Bjerknes, 20 June, 12 August 1902, BFC. The chair in question had been vacated by Boltzmann. Lorentz was asked, but declined. Wiener told Bjerknes that responses were expected soon from Paul Drude and Wilhelm Wien; he expected them to decline because they would want to continue as experimental physicists. Although Wiener suggested that Bjerknes might be next on the list, the chair went to Theodore Des Coudres. Bjerknes's nationality may have played a role in eliminating him from such an important professorship.

65. V. Bjerknes to Lorentz, 4 September 1904. HAL.

66. Ibid.; on the Swedish physics discipline during the early 1900s, see R. M. Friedman, "Nobel Physics Prize in Perspective," *Nature* 292 (1981), 793–98; Elisabeth Crawford and R. M. Friedman, "The Prizes in Physics and Chemistry in the Context of Swedish Science," in *Science, Technology and Society in the Time of Alfred Nobel*, ed. C. G. Bernhard, Elisabeth Crawford, and Per Sörbom (Oxford, 1982), pp. 311–31; Elisabeth Crawford, *The Beginnings of the Nobel Institution: The Science Prizes, 1901–1915* (New York and Paris, 1984), chap. 2. Bjerknes discusses mathematical physics in *Stockholms högskola 1878–1898*, pp. 68–69 and V. Bjerknes to J. A. Bonnevie, 8 May 1893, BFC.

67. V. Bjerknes to Lorentz, 4 September 1904, HAL.

proceedings of the Royal Swedish Academy of Sciences, he had to submit through a member. Only a limited number of articles could be published each year. What was worse, the proceedings were not widely read abroad. Bjerknes frequently complained that his works were buried in these proceedings; occasionally he managed to have them republished in foreign journals.[68]

Bjerknes saw, however, one opportunity to attain both greater prestige and a chance to advance his notion of the methods and direction for physics. The establishment in 1900 of the Nobel Committee for Physics and the promise of research institutions (to evaluate nominations for the prizes in chemistry, physics, and physiology/medicine) offered Swedish scientists opportunities for increasing their material resources and international prestige. Like many of his colleagues, Bjerknes believed the Nobel prize would transform Stockholm into a major center for international science, for the annual winners would spend a month or longer in Stockholm, and well-equipped, internationally oriented Nobel research institutions would soon be erected.[69] Initially the physics committee included no mathematical physicist, a deficiency that became apparent when members of the committee attempted in 1902 to evaluate Lorentz's contributions to electron theory. Arrhenius, one of the five physics committee members, began nominating Bjerknes for membership in 1902 in the hope that his colleague and friend could add expertise and help represent the högskola against the dominance of Uppsala physicists on the committee. Expecting to benefit from this position, Bjerknes turned down offers of help in returning to Norway, where he otherwise very much wanted to live. The extra financial benefit of being a member of the committee would increase his annual salary by a third. The authority to help define physics' priorities, especially at home, that came with this position seemed evident. After Arrhenius successfully obtained chemistry prizes for himself and for J. H. van't Hoff, he was about to obtain his own Nobel institute for physical chemistry. A Nobel physics institute was still to come.

Bjerknes's optimism both about election to the committee and about Stockholm's rapid emergence as a major research center proved ill-founded. Bjerknes's candidature, being closely associated with Arrhenius and with the högskola, became ensnared in personal and institutional rivalries and met with repeated defeat. According to

68. V. Bjerknes to C. A. Bjerknes, 28 May 1899; V. Bjerknes to Carl Runge, 9 December 1903, DM.

69. V. Bjerknes to Nansen, 18 January 1902, FNP; Arrhenius to Hjelt, 26 December 1898, 18 January 1900. EHP; Crawford and Friedman, "Prizes in Physics and Chemistry," 321–322; Crawford, *Beginnings of the Nobel Institution*, p. 177ff.

Arrhenius, the Royal Swedish Academy had made a foolish error in the eyes of those German and Danish physicists with whom he discussed the matter, for only in the academy and among Uppsala scientists would the Uppsala physicist Gustaf Granqvist be considered more worthy than Bjerknes.[70] No matter how Arrhenius tried to rectify the situation, little could be done. When Norway declared independence from the Swedish king in 1905, Bjerknes's candidature could expect even less sympathy. Bjerknes, feeling deeply hurt and insulted, sarcastically expressed his frustration: when return to Norway becomes possible, "then I'll only be happy for the opposition here and thank my friends for allowing me to get out without coming into contact with Nobel corruption."[71] If he could not count on support in his local milieu, how could he conceive of exerting influence abroad?

Despite his dismal situation in the world of physics, Bjerknes returned again and again to the problem of formulating a common set of equations for electromagnetic and hydrodynamic fields of force. Each step forward allowed him to indulge in momentary delight, but it also underscored his dilemma in physics. Each small triumph made Bjerknes realize the professional hollowness of the intellectual victory if he ever were to bring his and his father's work to completion. Upon making a breakthrough, he noted bitterly that he had no need to rush to publish; after all, in this "age of sensational physics," who cared?[72] Although always hoping to get other physicists to take up his research program, he never lost the feeling that he was working "in absolute loneliness."[73] As Bjerknes regarded the situation, he could not even command the attention of those whom he wanted to convince. Once, the chance occurred; his reaction is indicative of his frustration: "I have been attacked in *Annalen der Physik!* For the first time in my life attacked! I hope this signals that the dead silence which until now has surrounded my father's and my work finally has been broken. I am looking forward quite immensely to writing my answer. Luckily it's no fool I have against me. And the particular misunderstandings underlying his attack are easy to clear up, but behind the details there is a conflict on principles—and that is delightful."[74]

Unwilling to abandon a classical mechanical world view and not convinced that the new directions in physics would ultimately prove

70. Arrhenius to Knut Ångström, 2, 5 February, 2, 8 March 1904. K. Ångström Papers, KVA-SUB.

71. V. Bjerknes to H. Bjerknes, 26 February 1904, BFC.

72. V. Bjerknes to Lenard, 5 April 1904, PLP.

73. V. Bjerknes to Arrhenius, 23 December 1907, SAP; see ibid., 5 November 1910, SAP; V. Bjerknes to Nansen, 29 August 1912, FNP.

74. V. Bjerknes to Àrrhenius, 23 October 1909, SAP. Hans Witte wrote the article; Bjerknes's two replies appeared in *Annalen der Physik* in 1910.

fruitful, Bjerknes adopted the attitude that physicists would eventually return to the problem of the mechanical foundations of electromagnetism. He also admitted that the choice of underlying foundations for physical theory was mostly a matter of style within the
community of scientists.[75] Just as the pendulum of opinion swung
frequently between concepts of physical space as empty or full of
matter, so too the acceptance or rejection of mechanics as a basis for
physics would continue to fall in and out of favor. His view of how
physics develops gave him the conviction that at the very least a scientist must strive to draw his colleagues' attention and attempt to persuade them. As Bjerknes read the history of physics, certain underlying theoretical assumptions seemed simply to have been decided
upon by the community of researchers; the choice and definition of
problems and research programs was likewise open to influence. Little wonder that Bjerknes, having experienced the chaotic ferment in
fin-de-siècle physics, should find comfort in an attitude toward scientific advance which would permit him to hope that his research program, and Hertz's *Mechanics,* might yet become influential. His scientific career and professional status, however, could not depend upon
the future tastes of physicists. After a short sojourn in the position of
a promising young man on the way up in the 1890s, Bjerknes, by the
middle of the first decade of the new century, found himself headed
toward long-term residence in a state of scientific obscurity.

75. V. Bjerknes, "Vid. selvbekj."; idem, "Det nye optiske fænomen og Einsteins
relativitetsteori," *Naturen* 44 (1920), 161–86; idem, "Det mekaniske verdensbilde," ibid.
49 (1925), 321–39.

[2]

The Turn to Atmospheric Science: Contingency and Necessity

IT WAS from his Stockholm colleagues that Bjerknes first learned his physics was applicable to the study of atmospheric and oceanic motions.[1] He had been presenting the progress of his hydrodynamic investigations at meetings of the Stockholm Physics Society (Fysiska sällskapet). When he first presented his circulation theorem in 1897 he did not mention any potential application other than to the hydrodynamic analogies to electromagnetic phenomena.[2] Bjerknes had virtually no interest, and certainly no special competence, in meteorology and oceanography. Nils Ekholm and Svante Arrhenius, among others, frequently discussed geophysical topics at the Physics Society, and Bjerknes showed little interest. Then in February 1898, Bjerknes presented to the Physics Society a new lecture

1. For background on Stockholm science during this period, see *Stockholms högskola 1878–1898: Berättelse öfver högskolans utveckling under hennes första tjugoårsperiod, på uppdrag of hennes lärareråd utgifven af högskolans rektor* (Stockholm, 1899), esp. pp. 1–34, 67–103; *Stockholms högskola under åren 1899–1906: Berättelse, afgifven af Läraderådet* (Stockholm, 1907), esp. pp. 3–13, 26–32, 39–60; Fredrik Bedoire and Per Thullberg, *Stockholms universitets historia 1878–1978* (Stockholm, 1978), pp. 16–36. An overview of natural science in Sweden at the time can be found in Gunnar Ericksson, *Kartläggarna: Naturvetenskapens tillväxt och tillämpningar i det industriella genombrottets Sverige 1870–1914,* Umeå Studies in the Humanities 15 (Umeå, 1978).

2. Stockholm Physics Society, "Protokoll," held at Theoretical Physics Institute, University of Stockholm. The secretary or another leading member wrote detailed summaries of the biweekly meetings, which were published in part or in entirety in the major Stockholm newspapers. (References to these summaries are cited by newspaper and date.) The Physics Society was founded in 1891 to serve as a meeting place for Stockholm scientists wishing to follow developments in the physical sciences in Sweden and abroad; *Stockholms högskola 1878–1898,* pp. 80–81.

on the circulation theorem and this time suggested several applications to atmospheric and oceanic phenomena in which Stockholm researchers were particularly interested.[3] At a yet later presentation, in October 1898, his schematic application of the theorem to account for diverse geophysical phenomena reinforced the belief among the society's members that major advances could result from use of the theorem. Bjerknes published at this time the first two articles on the circulation theorem showing its potential value, respectively, for studying the analogies between hydrodynamic and electromagnetic fields and for comprehending the mechanics of atmospheric and oceanic motions.[4] Not only did the Stockholm context prove crucial for Bjerknes's initial contact with meteorology and oceanography; it also figured significantly in his further, sometimes almost reluctant involvement in geophysical studies.

Balloons and Herring in Sweden

During the 1890s some Stockholm researchers endeavored to apply the methods and laws of physics and chemistry to comprehending the macrophenomenal world. They hoped to bring the erratic and seemingly random phenomena of the seas, atmosphere, and solid earth into the domain of exact physical science, to establish a "cosmical physics." Arrhenius, the högskola's professor of physics and Bjerknes's only close friend in Stockholm, had begun to focus almost exclusively on this field.[5] In part he strove to apply his theory of ionic solutions to diverse phenomena outside physical chemistry proper; in part he considered a general synthesis of terrestrial and astrophysical phenomena an effort worthy of his full attention. He encouraged Bjerknes to devote attention to this field and put him in contact with other researchers. Some of these Stockholm investigators immediately recognized in Bjerknes's circulation theorem a potential means of resolving problems that in 1897 had become especially well defined. In this respect, Bjerknes's entry into geophysical research was largely a consequence of Ekholm's interest in cyclone theory and ballooning and of Pettersson's interest in oceanography and fishery.

Ekholm's enthusiasm for Bjerknes's circulation theorem stemmed

3. *SD*, 28 February 1898, and Ab, 2 March 1898.
4. V. Bjerknes "Ueber die Bildung von Cirkulationsbewegungen und Wirbeln in reibungslosen Flüssigkeiten," *Skrifter udgit af Videnskapsselskabet i Christiania*, Math.-natural sciences sec., 1, no. 5 (1898); idem, "Ueber einen hydrodynamischen Fundamentalsatz und seine Anwendung besonders auf die Mekanik der Atmosphäre und des Weltmeers," *VAH*, n.s. 31, no. 4 (1898/99).
5. Svante Arrhenius to Edward Hjelt, 18 January 1900, EHP.

from his prior work on cyclone theory and, more immediately, from his preoccupation with the Swedish balloon expedition to the North Pole. Ekholm studied meteorology in Uppsala with Sweden's first professor in this subject, H. H. Hildebrandsson. After serving as an assistant at Uppsala University's meteorological observatory, he obtained in 1890 a similar post with the Central Bureau of Meteorology in Stockholm. Although his investigations covered a broad range of topics including instrument design, cloud studies, cyclone theory, and methods for weather forecasting, he was especially attracted to problems related to ballooning and polar exploration. He led the 1882–83 Swedish meteorological expedition to Spitsbergen, was a member of the aborted balloon expedition to the North Pole in 1896, and was a key figure in planning the repeat effort, in 1897, to place the national flag on the pole.

The polar effort was part of a Swedish and Norwegian tradition of polar exploration grounded in scientific study and nationalist enthusiasm. A. E. Nordenskiöld's 1878–80 voyage on the *Vega* through the Northeast Passage around Asia is, of course, one of the most famous examples. After Fridtjof Nansen's Norwegian expedition on the ship *Fram* to reach the North Pole captured the world's attention in the early 1890s, Nordenskiöld, in 1894, suggested to engineer and ballooning enthusiast S. A. Andrée the idea of reaching the North Pole by balloon.[6] With financial assistance from, among others, King Oscar II and Alfred Nobel, Andrée and ballooning companion Ekholm began preparations for "perhaps the greatest balloon adventure of all time."[7] In the tradition of Nordenskiöld and Nansen, the expedition was to be scientific. The Stockholm Physics Society, Arrhenius's physics institute, and the Swedish Society for Anthropology and Geography, among other institutions, planned the scientific observations, experiments, and instrumentation for the trip. In August 1896, Andrée and Ekholm were joined by Bjerknes's assistant Nils Strindberg for the first attempt to launch the balloon from Spitsbergen. Although Ekholm tried to determine the best possible time for the voyage, bad weather forced a cancellation. Ekholm helped plan a second attempt for July 1897 but did not this time participate in the flight. The balloon with its crew of Andrée, Knut Frankæl, and Strindberg disappeared.

6. S. A. Andrée, Nils Strindberg, and Knut Fraenkel, *Med örnen mot polen: Andrées polarexpedition år 1897*, ed. Swedish Society for Anthropology and Geography (1930; rpt. Uddevalla, 1978), pp. 27–29ff.

7. Charles Harvey Gibb-Smith, *A Brief History of Flying: From Myth to Space Travel* (London, 1967), p. 24.

Ekholm's keen interest in ballooning naturally interested him in the motions of the air aloft, where balloons have to navigate.[8] This little-explored and scarcely understood subject became critical when, during fall 1897, he, Arrhenius, and other scientists considered a rescue mission. They had to decide, from a few weather observations taken at ground level, where the upper-air currents might have taken the balloon. No reliable theoretical guidelines for inferring probable three-dimensional atmospheric motions were at hand[9]—hence Ekholm's immediate interest in Bjerknes's theorem and his alleged exclamation that now it might be possible to add a vertical dimension to weather maps.[10]

Ekholm's interest was conditioned by his own earlier meteorological investigations. He had shown that the atmosphere at times possesses characteristics similar to those of the fluid postulated by Bjerknes for his circulation theorem. Ekholm had shown in 1891 that the air in the vicinity of cyclones reveals an incongruity between pressure and density; that is, on a horizontal projection near the surface, lines drawn of equal values of pressure and density do not always coincide.[11] Ekholm was convinced this unexpected distribution of pressure and density must be a factor in the formation of cyclones. Similarly, Bjerknes's initial geometric interpretation of the circulation theorem entailed three-dimensional surfaces of equal pressure and density intersecting to form a series of tubes or, as he called them, solenoids. Because of the tendency toward rotation in each rectangular solenoid in the crisscross lattice of intersecting surfaces, the occurrence of a skewed distribution of pressure and density in the atmosphere should result in circulation, the direction and intensity of which in principle could be determined by Bjerknes's theorem. According to this model, the rate of increase of circulation in a closed curve is related to the number of solenoids encompassed by the curve. In a situation in which surfaces of pressure and density coincide, no solenoids exist and the circulation of a fluid curve remains invariable with time, as in the case of the classical theorems for ideal fluids. Ekholm's charts showed only a two-dimensional projection near the earth's surface, but they nevertheless suggested to him a likelihood that Bjerknes's theorem could be useful for

8. Ekholm was an active aeronautics enthusiast. Having flown with Andrée in the early 1890s, he worked closely with all ballooning and, later, other aeronautical activities in Sweden and helped found and lead the Swedish Aeronautical Society in 1900; *Jubileumsårsbok för Kungliga Svenska aeroklubben 1931* (Stockholm, 1931), pp. 17–23.

9. Nils Ekholm, "L'Expëdition polaire en ballon de m. S.-A. Andrée," *L'aerophile* 6 (1898), 10, 21–23.

10. V. Bjerknes to C. A. Bjerknes, 7 March 1898, BFC; Bjerknes recalls Ekholm's reaction in a letter to V. Walfrid Ekman, [?] January, 16 February 1946, VWE.

11. Nils Ekholm, "Etude des conditions météorologiques à l'aide des cartes synoptiques représentants la densité de l'air," *Bihang VAH* 16 (1891), sec. I, no. 5.

studying three-dimensional atmospheric motions and for comprehending the nature of cyclones.[12]

Consequently, when Bjerknes discussed his circulation theorem at the Physics Society in February 1898, he remarked that the recent greater availability of observations from higher levels, made possible by balloons and kites, might facilitate use of the circulation theorem to test Ekholm's theory that the air currents in a cyclone arise from unequal distribution of density and pressure. Bjerknes continued by discussing, based on the circulation theorem, the possible motions of Andrée's balloon within an Arctic cyclone. Subsequently, Ekholm told Bjerknes that the theorem could lead to major innovations—at least methodological ones—in meteorology and spent a full day with him reviewing current meteorological literature. Bjerknes also learned that Arrhenius wanted to use the theorem as a basis for a chapter on the mechanics of the atmosphere and oceans in his forthcoming book on cosmic physics.[13]

Bjerknes learned from Ekholm that the circulation theorem might be able to resolve a conflict on the nature of extratropical cyclones (which are midlatitude low-pressure systems).[14] For almost twenty years uncertainty had prevailed as to whether cyclones are formed by rising currents of warm air, which release thermal energy during condensation, or by mixing of different air currents. Bjerknes proposed constructing a vertical profile of surfaces of equal pressure and density (or specific volume) using data obtained from recording instruments carried aloft by kites. It might then be possible to calculate whether the circulation produced by the unequal distribution of density—as manifested by the presence of solenoids—could account for the cyclone's energy. If the circulation within the cyclone did arise from the presence of solenoids, the various theories on the origin of cyclones could be tested by their ability to yield the observed differential distribution of pressure and density.

In response, the Physics Society enthusiastically established a committee to construct kites or "flying machines driven by electric motors" to obtain data.[15] Pettersson offered to pay most of the initial

12. *Ab,* 11 October 1897; V. Bjerknes to C. A. Bjerknes, 7 March 1898, BFC.

13. V. Bjerknes to C. A. Bjerknes, 7 March 1898, BFC. The chapter for Arrhenius's book, *Lehrbuch der kosmischen Physik,* vol. 1 (Leipzig, 1903), was eventually completed by Bjerknes's assistant, J. W. Sandström.

14. Bjerknes, "Ueber einen hyd. Fundamentalsatz und seine Anwendung," pp. 32–33; for an in-depth discussion of the cyclone controversy, see Gisela Kutzbach, *The Thermal Theory of Cyclones: A History of Meteorological Thought in the Nineteenth Century* (Boston, 1979), esp. pp. 63–206.

15. *SD,* 19 October 1898; V. Bjerknes to C. A. Bjerknes, 22 October, 30 November 1898, BFC; V. Bjerknes to Hugo H. Hildebrandsson 12 October 1898, HHH.

expenses; he and Arrhenius subsequently obtained funds for the kites and recording instruments from the Lars Hiertas Memorial Fund. When at this time a very bright and energetic but poorly educated young man from northern Sweden was sent by his village and local sawmill to the Stockholm Högskola to develope his talents, the faculty directed this man, J. W. Sandström, to Bjerknes. Pettersson and Arrhenius obtained a fellowship for Sandström to allow him a chance to calculate an example of the solenoids present in a cyclone. In spite of his commitment at that time to the rapid completion of the book based on his father's studies, Bjerknes could not turn his back on the cyclone investigation. After all, the kite and flying-machine committee had been established "in his honor": he had no choice but to assist with the study of kites and with Sandström's graphic calculation of the solenoids in a cyclone.[16] Bjerknes quickly realized he could not avoid "getting sucked into the [international] meteorological vortex".[17]

Bjerknes and Sandström React

As soon as his geophysical work became known outside Scandinavia, Bjerknes was pressured to devote more time to these studies. Two leading meteorologists, the American Cleveland Abbe and the Austrian Julius Hann, turned to Bjerknes for contributions to the important meteorological journals they respectively edited, *Monthly Weather Review* and *Meteorologische Zeitschrift*. As a long-time active participant in the debate over the nature of cyclones, Hann hoped that Bjerknes's theorem could help to resolve that debate. For his part, Abbe was enthusiastic about any measure that would help introduce rigorous physics and mathematics into meteorology. Abbe had tried unsuccessfully to institutionalize meteorology as an academic discipline as part of a broader "terrestrial physics"; he was still determined to advance meteorology's claim to be a legitimate, rigorous science.[18] Although he first declined Abbe's offer because he was preoccupied with his book, Bjerknes did begin a correspondence with Abbe that proved valuable during the next several years. Abbe helped arrange for Bjerknes to receive a unique set of upper-air data obtained with kites during the passage of a cyclone and an anticyclone (a

16. V. Bjerknes to C. A. Bjerknes, 1 January 1899, BFC. The kites provided a refreshing diversion.

17. V. Bjerknes to Honoria Bjerknes, 25 June 1899, BFC.

18. Ibid.; V. Bjerknes to C. A. Bjerknes, 1 January, 19 April 1899, BFC. Cleveland Abbe to V. Bjerknes, 9 December 1898, 11 January 1899; Julius Hann to V. Bjerknes, July 1899, all VBP. Pettersson alerted Hann to Bjerknes's study. On Abbe's attempts to institutionalize meteorology in academia, see M. A. Dennis, "Stormy Weather: Cleveland Abbe, American Meteorology, and the Failure of a Disciplinary Project, 1869–1915" (seminar paper, Dept. of History of Science, Johns Hopkins Univ.).

high-pressure system) over the Blue Hill Observatory in Massachusetts. Using these data Sandström constructed a three-dimensional cross section showing surfaces for equal values of pressure and density and concluded, with Bjerknes, that enough solenoids were present to produce, within several hours, even the strongest winds observed.[19] Of course these data were rather incomplete and not synchronized. Moreover, the theorem did not include terms for the effects of friction and the earth's rotation. Still, this first attempt to apply the theorem to a set of observations convinced Bjerknes that he had a new "rational dynamic principle" for comprehending atmospheric motions.[20]

In response to this suggestive finding, Bjerknes returned to the circulation theorem and analyzed it from a geophysical rather than a purely hydrodynamic perspective. In "Circulation Relative to the Earth," he expanded the circulation theorem to its now-classic formulation, including terms for forces arising from the earth's rotation and, in approximate form, from friction.[21] Sandström applied this result to an additional set of data sent by Abbe, after which he proposed a solution to the cyclone debate. The controversy was complicated by observations indicating that cyclones seemed to possess at times either a cold or a warm core. Sandström's application of the circulation theorem led him to conclude, somewhat convolutedly, that the distribution of solenoids changed with time, making possible an evolution from a warm to a cold center within a cyclone.[22] Naturally Bjerknes was glad that this result (unlike his physics) might attract attention, even controversy, but he continued to take little direct interest in the cyclone debate. For him, Sandström's result was important because it showed the advantage of using mechanics for comprehending atmospheric phenomena; details of the debate were largely secondary. Bjerknes's attention to and ultimate interest in meteorology came from a different quarter.

Pettersson and "Rational Fishery"

At the time of the cyclone developments, applications of the circulation theorem to oceanographic problems underscored for Bjerknes the theorem's value for geophysical research. Pettersson, professor of chemistry at the Stockholm Högskola, believed in the need to apply precise analytic

19. Reported in *SD*, 15 May 1899, and *DN*, 16 May 1899.

20. V. Bjerknes, "Das dynamische Princip der Zirkulationsbewegungen in der Atmosphäre," *MZ* 17 (1900), 153.

21. V. Bjerknes, "Cirkulation relativ zu der Erde," *VAÖ* 58 (1901), 739–57 and *MZ* 19 (1902), 97–108.

22. See Kutzbach, *Thermal Theory of Cyclones*, pp. 162–80, for a discussion of Sandström and Bjerknes's work on cyclone theory, 1899–1902, in the context of the debate.

and physical techniques to the study of the oceans. His entry into oceanographic work was very much a Swedish affair. First, Pettersson analyzed the oceanographic observations taken on Nordenskiöld's *Vega* expedition. Then, several years later, he became fully committed to studying the oceans. In 1890 the Royal Swedish Academy of Sciences asked Pettersson to complete the chemical analysis of observations taken in the 1870s during the first state-supported oceanographic survey of the Baltic Sea, Skagerak, and Kattegat.

Meanwhile, Pettersson's friend and colleague from chemical studies at Wiesbaden, Gustav Ekman, was completing analyses of samples obtained during the winters of 1878 and 1879 off the Swedish west coast (Bohus County). These surveys had been prompted by the sudden return of coastal herring after a seventy-year absence that had left the local society in economic ruin. Ekman's findings tended to justify the public and private expenditures for such studies. The economically and socially important herring seemed to follow a specific underwater stratum of particularly high salinity and temperature. When this layer was present in the Bohus County coastal waters and fjords, so too were the herring; when it disappeared, so did the fish— and the local prosperity.

Pettersson concluded from Ekman's finding that understanding and being able to predict the sea's internal motions could aid fishery. Pettersson and Ekman organized a new series of surveying expeditions to construct comparative three-dimensional cross sections of the seas surrounding Sweden. They published the first seasonal profiles of the Skagerak and Kattegat in the spring of 1897, just when Bjerknes formulated his circulation theorem. These charts showed the changing occurrence of the specific stratum that the herring follow. The need to understand the dynamics of this stratum, and of sea currents in general, had been brought home by a sharply decreased catch as the fish failed to appear close to shore after the 1895–96 season.[23] Pettersson considered the identification, tracking, and eventual prediction of this layer's motions a major scientific and social challenge.[24]

Pettersson immediately embraced Bjerknes's discovery as the theoretical breakthrough necessary to establish an oceanographic science with which fishery could be planned. According to Bjerknes's the-

23. "Sillfiske," *Nordiska familjebok* (Stockholm, 1917), 25:527–28.

24. On Pettersson and contemporary Swedish oceanography: Otto Pettersson, *En Självbiografi* (Göteborg, 1938); V. W. Ekman, "Otto Pettersson 12/2 1848 to 17/1 1941," *Svensk geografisk årsbok* 17 (1941), 85–92; A. [*sic*] Pettersson, "Den internationella utforskningen af de nordiska hafven," *Ymer* (1904), 94–112; Otto Pettersson and G. Ekman, "De hydrografiska förändringarna inom Nordsjöns och östersjöns område under tiden, 1893–1897," *VAH*, n.s. 29, no. 5 (1897).

orem, the configuration of a distinct stratum of water possessing salinity and temperature differing from the surrounding sea ought to result in the formation of circulation and hence of internal motions. At the small högskola, Pettersson and Bjerknes certainly had opportunity to exchange ideas; their mutual friend Arrhenius would surely have brought them together. In his February 1898 lecture at the Physics Society, Bjerknes used Ekman and Pettersson's charts of the Skagerak as the basis for a general scheme for comprehending ocean currents by means of the circulation theorem. Pettersson was now convinced of the feasibility of a "rational fishery." His use of this term indicates both a desire to base all aspects of fishery on scientific principles and methods, and a belief that scientists had the key to an efficient, profitable fishery. His far-reaching, almost fantastical vision included the creation of exact sciences of the oceans and the atmosphere that could yield predictions of the meandering of ocean currents. His firm belief that the oceans and atmosphere interact meant that long-term weather forecasts as well as predictions of motions in the oceans could only be achieved by studying the two systems together. Bjerknes's circulation theorem was to be the cornerstone for both new sciences: Pettersson began supporting—and directing—Bjerknes by drawing him into international projects and by helping him obtain funds for assistants and experiments.[25] Indeed, Pettersson was already taking steps to secure an organizational foundation for his rational fishery.

When trying to make surveys of the seas around Sweden, Pettersson had recognized the need for international cooperation. To bring about regular surveys of the seas around northern Europe as a first step toward this goal, Pettersson convinced King Oscar II of the project's value to Sweden. The king agreed to convene an international conference in Stockholm.[26] Although the main task of this 1899 conference was to plan international cooperation, Pettersson persuaded Bjerknes to leave his physics work for a while and present his preliminary geophysical thoughts to the assembled scientists.[27] A year later, Pettersson arranged in Gothenburg another gathering that was to be a preparatory meeting for the next official international conference. Bjerknes was again a special guest lecturer on the potential use of his theorem to introduce exact methods into meteorology and oceano-

25. *SD*, 28 February 1898, and *Ab*, 2 March 1898.

26. Pettersson, *Självbiografi*, pp. 9–11; A. E. J. Went, *Seventy Years Agrowing: A History of the International Council for the Exploration of the Sea* (Copenhagen, 1972), pp. 3–16; Susan Schlee, *The Edge of an Unfamiliar World: A History of Oceanography* (New York, 1973), pp. 206–13.

27. V. Bjerknes to C. A. Bjerknes, 21 October 1899, BFC.

graphy.[28] Bjerknes and Sandström also demonstrated the feasibility of collecting upper-air data with kites from sea-going oceanographic vessels. Acting on Pettersson's suggestion, the assembled scientists agreed that future oceanographic surveying expeditions must endeavor to collect data from both the sea and the air so as to gain insight into the interactions between the two fluids. Bjerknes was asked to provide suggestions for this work, a request that caused him once again to reflect on the application of his theorem.

This growing interest among northern European nations in the scientific study of the oceans was yet another step in drawing Bjerknes's attention to the study of oceans and atmosphere. When his Norwegian colleague Fridtjof Nansen read the 1898 articles on the circulation theorem, he wrote to Bjerknes almost immediately (Fig. 1).[29] Nansen was then analyzing oceanographic observations taken during his epoch-making voyage on the *Fram* in the Arctic. Now best known as an explorer, statesman, and humanitarian, Nansen also held a professorship at the university in Christiania and occupied himself with various oceanographic projects.[30] In his note to Bjerknes, Nansen inquired whether the circulation theorem could help explain the mysterious dead-water phenomenon: ships, including the *Fram,* would suddenly become stuck in a fjord or bay and lose response to steering. Bjerknes responded quickly to his famous countryman.[31] He suggested, following Nansen's insight, that internal waves formed on the boundary surface separating strata of differing density might produce this effect. To explore the issue further Bjerknes asked an Uppsala student who was following his lectures, V. Walfrid Ekman, to investigate the problem both experimentally and theoretically. Bjerknes arranged for work space and supervised Ekman's investigations.[32]

Then, in 1900, while visiting Stockholm, Nansen posed another problem for Bjerknes.[33] Observations from Nansen's voyage indi-

28. Ibid., 30 September, 4 November, 3 December 1900, BFC. Secondary literature on the history of international oceanographic cooperation seems to have forgotten Pettersson's second conference; for a summary, see *Göteborgs handel- och sjöfartstidning,* 21 November 1900, and *Ab,* 29 November 1900.

29. Fridtjof Nansen to V. Bjerknes, 21 October 1898, VBP.

30. Nansen's scientific work has yet to be studied in depth. Steinar Kjærheim's monumental editorial work on Nansen's correspondence reveals Nansen's dedication to oceanography; see esp. his *Fridtjof Nansen Brev,* vol. 2, 1896–1905 (Trondhjem, 1961) and idem, "Nansen som forskningspolitiker," *Forskningsnytt fra Norges almenvitenskapelige forskningsråd* 21, no. 7 (1976), 29–35.

31. V. Bjerknes to Nansen, 31 October 1898, FNP.

32. *Stockholms högskola under åren 1899–1906,* p. 28; Bjerknes to Nansen, 6 April 1900, FNP; *Ab,* 14 April 1900.

33. Nansen to V. Bjerknes, 11, 17, 19 December 1900, VBP. Nansen and Bjerknes corresponded frequently on oceanographic issues in 1901.

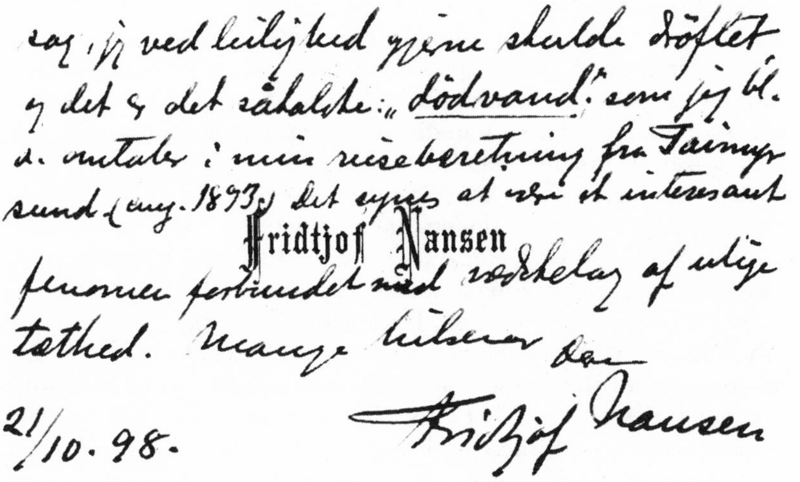
sag, jeg ved leilighed gjerne skulde drøftet og det er det såkaldte „dødvand" som jeg bl. a. omtaler i min reiseberetning fra Taimyr sund (aug. 1893). Det synes at være et interessant fenomen forbundet med vekslag af ulige tæthed. Mange hilsener fra
Fridtjof Nansen
21/10.98.
Fridtjof Nansen

Fig. 1. Nansen's note to Bjerknes concerning "dead water." Nansen was writing to thank Bjerknes for a copy of "Ueber einen hydrodynamischen Fundamentalsatz . . . ," which arrived just as he was working on the *Fram* expedition's hydrographic results. He expresses a desire to discuss with Bjerknes the problem of "dead water," which Nansen had written about in his travel account. Calling the phenomenon interesting, Nansen assumes it arises in connection with fluid layers of differing density. From the Fridtjof Nansen file, VBP. Courtesy Oslo University Library.

cated that both ship and free-floating ice had drifted at an angle between 20° and 40° to the right of the wind, rather than with the wind according to established theory. Bjerknes again called on Ekman, who was still working on the dead-water problem. In a matter of hours Ekman produced a preliminary solution. When he reworked the problem in light of Bjerknes's expanded circulation theorem that took into account friction and the earth's rotation, Ekman established the now-classic mathematical derivation of the three-dimensional structure of wind-driven currents on a rotating earth, called the Ekman spiral. This result broke with long-held traditions in oceanography and, along with the dead-water investigation, proved impressively the circulation theorem's value for comprehending motions in the ocean.[34]

34. Nansen had asked Bjerknes (17 December 1900, VBP) whether Ekman could work with him in Christiania. When Nansen visited Bjerknes in Stockholm on 5 November 1901, the problem was turned over to Ekman, who produced a first solution that night and continued intensive discussions with Nansen until the latter left on the ninth. (Ekman later copied an extract from his diary onto a letter from Bjerknes, 28 March 1926, VWE). For Bjerknes's enthusiasm and Ekman's first results, see V.

When the international group assembled by Pettersson established itself in 1902 as the International Commission for the Exploration of the Sea, it created an international laboratory in Christiania under Nansen's supervision. Ekman became the laboratory's first assistant and, soon thereafter, assistant director. Hoping meanwhile to develop oceanography further in Norway, Nansen in 1901 sent Bjørn Helland-Hansen to Bjerknes.[35] After learning the theoretical and practical applications of the circulation theorem from Bjerknes and Sandström respectively, Helland-Hansen returned to Norway and began oceanographic investigations in his position at the newly founded Fishery Directorate in Bergen. Under the leadership of the zoologist Johan Hjort, whose belief in studying the physical properties of the seas to benefit fishery resembled Pettersson's, the Fishery Directorate began a series of surveying expeditions in the North Sea. Helland-Hansen worked with these and with international expeditions in the early 1900s. Using Bjerknes's circulation theorem, he, sometimes with Sandström, devised innovative analytic methods of calculating stationary ocean currents. Ekman, Helland-Hansen, and Sandström, using the circulation theorem, managed to produce the cornerstones of twentieth-century dynamic oceanography.

Professional Opportunities and Pitfalls in Meteorology

Bjerknes was pleased. Physics had, of course, been applied to the atmosphere and ocean during the nineteenth century. Atmospheric thermodynamics had already developed special techniques and problems that were distinctly meteorological rather than laboratory-experimental.[36] Applications of classical hydrodynamics to idealized geophysical fluids could also be noted. But these laws and theories had been applied separately from each other, so that thermal energy had not been related to motions within these fluids. In Bjerknes's

Bjerknes to C. A. Bjerknes, 1 December 1901, BFC. Ekman's first publication, "Om jordrotationens inverkan på vindströmmer i hafvet," *Nyt Magazin for naturvidenskaberne* 40 (1902), 1–22, did not attract much attention; his more comprehensive treatment published three years later became a classic, "On the Influence of the Earth's Rotation on Ocean-Currents," *AMAE* 2, no. 11 (1905) 1–53. The older, established school of thought, which included the Norwegian meteorologist H. Mohn and several Uppsala meteorologists, immediately opposed the Ekman results; see, e.g., V. Bjerknes to Ekman, 15 March 1903, VWE; Mohn to Hildebrandsson, 24 January, 12 October 1904, 13 February 1908. HHH. Such younger exponents of a new oceanography as Nansen, Pettersson, Bjørn Helland-Hansen, and Johan Hjort welcomed these results.

35. Nansen to V. Bjerknes, 12 March 1901, VBP; V. Bjerknes to C. A. Bjerknes, 1 December 1901, BFC.

36. Elizabeth Garber, "Thermodynamics and Meteorology (1850–1900)," *AS* 33 (1976), 51–65.

fluid the important variations of density that do not depend on pressure and that lead to circulation arise chiefly from differences in temperature and humidity in the atmosphere and differences in temperature and salinity in the ocean. Therefore, he asserted, the circulation theorem promised to elucidate real—not idealized—atmospheric and oceanic processes: "No matter how complicated the conditions are, we nevertheless always treat the real winds and their real causes. In this regard the present theory differs fundamentally from ordinary dynamic theories that are founded on the solution of special integrals of the equations of motion, and for which one must first assume a general, farfetched idealization of actual conditions before the theory can be brought to apply."[37] He was also pleased that his and his assistants' geophysical studies provided insights and methods that could be used in the continuing study of hydrodynamic analogies to electromagnetism.[38] It was, after all, the completion of the mechanical foundations of electromagnetism, and of the ether in general, that still held Bjerknes's primary allegiance and professional hopes.

Before I turn to Bjerknes's increasing involvement in geophysical study, let me digress to examine the issue of whether Bjerknes, had he not been in Stockholm, would have learned of the circulation theorem's applicability to studies of the atmosphere and ocean. Just ahead of Bjerknes, Ludwig Silberstein in Poland also arrived at a generalized circulation theorem.[39] Silberstein considered the theorem a formal generalization, for he doubted the existence of any fluid in which density is not a function of pressure only, except possibly for very short periods. Just as Bjerknes had once hesitated to question the reigning dogma of the conservation of circulation and vortex motions, Silberstein, in the face of Helmholtz's and Kelvin's results, tended to regard his own construction as an interesting curiosity without much physical significance. In fact, Bjerknes came to assume that many others had probably derived the same results math-

37. Bjerknes, "Dynamische Princip der Zirkulationsbeweg.," p. 153.

38. Bjerknes used interpretations of the circulation theorem developed for explaining geophysical phenomena ("Ueber einen hyd. Fundamentalsatz und seine Anwendung") to help him generalize his analogies: "Ueber Wirbelbildung in reibungslosen Flüssigkeiten mit Anwendungen auf die Analogie der hydrodynamischen Erscheinungen mit den elektrostatischen," *AMAE* 1 (1903), 225–50 and *Zeitscrift für Mathematik und Physik* 50 (1904), 422–43; Bjerknes's assistant from 1907 to 1910, Olaf Devik, claimed that at the time Bjerknes regarded the atmosphere "exactly as a big laboratory for hydrodynamics" (interview, January 1975, Oslo).

39. Ludwig Silberstein, "Ueber die Entstehung von Wirbelbewegungen in einer reibungslosen Flüssigkeit," *Bull. Int. de l'Académie des Sciences de Cracovie*, no. 6 (1896), 280–90.

ematically (the formal derivation required little brilliance) but that nobody else had understood their physical significance.[40] Because he was determined to extend the hydrodynamic analogies, Bjerknes was willing to break with prevailing thought and assert that the equations necessarily possessed some physical meaning. It was only after Ekholm and Pettersson drew attention to the theorem's implications for understanding atmospheric and oceanic phenomena that Bjerknes began to think about geophysical questions, and these men, with Nansen and others, kept him involved. It is in this regard that Stockholm, and more generally Scandinavia, was crucial to his involvement with meteorology and oceanography.

In the early 1900s, that involvement increased. Often Bjerknes divided his time equally between geophysical applications and his major preoccupation, the hydrodynamic analogies.[41] Bjerknes changed his priorities in two phases. Late in 1903 he formulated a major project for establishing an exact mechanical physics of the atmosphere that would allow "rational" precalculations of atmospheric conditions; in 1906 he first considered this project his main concern for the foreseeable future. An understanding of Bjerknes's reasoning and intentions helps make this meteorological research project and his future career choices comprehensible.

Bjerknes's conversion to a research program in geoscience was facilitated by the contrast that his early experiences in geophysics provided to his arduous, almost thankless endeavors in theoretical physics. From the very start, Bjerknes noted the differences in his ability to gain attention and exert authority. He began publishing in meteorological journals through the initiative of the editors and attracted reactions almost immediately. In contrast to the lack of discussion or debate over his hydrodynamic analogies, he and his assistants' new geophysical results almost always "create[d] a stir."[42] More significant, his training in mathematical physics allowed Bjerknes to assert scientific authority over the "old fogy authorities [*autoritetsstabeiser*] who look down their noses."[43] When one such meteorologist attacked him in *Meteorologische Zeitschrift* on the manner by which he defined various vector quantities, Bjerknes commented gleefully to his father: "His name is Müller, and he is supposedly an authority because he knows how to integrate. In his attack he has blundered so pitiably, that I can defend myself by going on the offensive, which after all is correct according to the rules of the art of war. And this will be an

40. V. Bjerknes to Ekman, 16 February 1947, VWE; Bjerknes, "Fra de pulserende kuler til polarfronten," *Ymer* 59 (1939), 190–92.
41. V. Bjerknes to Ekman, 20 October 1903, VWE.
42. V. Bjerknes to C. A. Bjerknes, 1 December 1901, BFC.
43. Ibid., 30 September 1900, BFC.

offensive he won't forget . . . so that in the future I can avoid such sudden attacks by scientific half-cultivated 'authorities.' "[44]

After meeting a number of German meteorologists, Bjerknes concluded that they respected him as a physicist showing an interest in their field but did not quite know what to make of his research. He wondered at times whether many meteorologists actually understood the circulation theorem and its applications.[45] But, over and beyond such musings, he appreciated that in Sweden and Norway, geophysical sciences, unlike theoretical physics, commanded a comparatively healthy amount of research support. Funding as well as important results seemed to come relatively easily; the initial attention and support prompted Bjerknes to comment, "This is a gratifying field in which to work."[46] His endeavors brought prestige and delightful rewards. Pettersson arranged impressive accommodations for him while he traveled to and attended the oceanography meeting in Gothenburg. Not only were the experiments with kite design fun and well funded, he even received special permission from King Oscar II to fly his kites on grounds belonging to the royal family.[47]

Bjerknes also recognized that in contrast to theoretical physics, he might be able to establish a school of disciples and assistants in meteorological and oceanographical research. Expectations were high both in Sweden and Norway that these sciences were about to experience significant local growth. Pettersson used his influence to expand the Swedish hydrographical research institute into the larger Swedish Hydrographic—biologic Commission, which organized several physical oceanographic and marine biologic projects. Results from these studies were published in high-quality format for international circulation. Pettersson began recruiting Sandström to these projects. In the meantime Nansen was trying to attract Sandström to the new meteorological observatory in Bergen, where he could then collaborate with Helland-Hansen. Although Ekman was unfairly passed over for a professorship at the university in Christiania, he was assuming most of the responsibilities for the international oceanographic laboratory. In short, all of Bjerknes's students were immediately in demand; in fact Bjerknes had to compete for Sandström's time.[48] Still, Bjerknes only gradually turned to this area of research as a profes-

44. Ibid.
45. V. Bjerknes to J. A. Bonnevie, 20 September 1903, BFC; V. Bjerknes to Carl Runge, 19 January 1903, DM.
46. V. Bjerknes to C. A. Bjerknes, 23 February 1901, BFC.
47. Ibid., 7 April 1901, BFC.
48. V. Bjerknes to Nansen, 4 September 1904, FNP; Nansen to cabinet minister W. A. Wexelsen, in Kjærheim, *Fridtjof Nansen Brev*, 3:229–31; V. Bjerknes to Ekman, 15 March 1903, VWE, discusses Mohn and O. Schiøtz's opposition to Ekman's work and candidacy.

sional alternative to the hydrodynamic actions at a distance, even after he understood the obstacles entailed in continuing the latter. This reluctance to commit his scientific career in this direction should not be surprising in light of the state of meteorology at that time.

Meteorology around 1900 was not a science to which leading physicists willingly devoted themselves. True, during the last quarter of the nineteenth century, physicists attempted on occasion to comprehend atmospheric phenomena with their theories and laws, sometimes with success. But those who spent too much energy and time in such endeavors soon found themselves in a quagmire. Bjerknes noted that leading physicists asked him rhetorically, "Why don't gifted young physicists go into meteorology?" When he visited the head of the Physikalisch-Technische Reichsanstalt in Charlottenburg (Berlin) and related his new interest in the mechanics of atmospheric phenomena, the elderly Friedrich Kohlrausch supplied a blunt answer: "A physicist who goes into meteorology is lost. So it went with Mascart. So it went with Bezold."[49]

Bjerknes learned quickly what his colleagues knew: a lack of scientific rigor was acceptable to most of the international meteorological community. Was this the kind of company that a self-respecting, status-conscious physicist would like to keep? Alternatively, with whom could Bjerknes keep company in theoretical physics? Some supportive physicists tried to convince him to devote time to his mechanical physics rather than to meteorology; Lenard, for example, claimed that he preferred to see Bjerknes's equations applied to the ether rather than in the air.[50] But fortunately for Bjerknes, the opportunity to effect major changes in meteorology as a science and a profession was just then beginning to emerge: the "conquest of the air" by aeronautics held out the hope that a comparable conquest of the air by physics might be feasible and worthy of attention.

Aerology benefited from advances in aeronautics.[51] Interest and

49. V. Bjerknes to Hugo Hergesell, 30 June 1926, copy, VBP; V. Bjerknes to C. W. Oseen, 16 July 1929, CWO. Bjerknes recalls the discussion with Kohlrausch as if it had just happened. E. E. Mascart and W. von Bezold held professorships in physics in France and Germany before becoming in the late nineteenth century the heads of their respective national weather services.

50. Philipp Lenard to V. Bjerknes, 7 September 1909, VBP.

51. For background on the early years of aerology, see Richard Assmann and Arthur Berson, *Wissenschaftliche Luftfahrten*, vols. 1–3 (Braunschweig, 1899–1900); H. Hoernes, *Die Luftschiffahrt der Gegenwart* (Vienna, Pest, Leipzig, 1903), pp. 90–180; Arthur Berson, "Aerologische Forschung," in *Buch des Fluges*, ed. H. Hoernes (Vienna, 1911), 1:529–74; Franz Linke, *Die Luftschiffahrt von Montgolfier bis Graf Zeppelin* (Berlin, n.d.), pp. 219–304; Hugo Hergesell, "The Work of the International Commission for Scientific Aeronautics," *Aeronautical Journal*, January 1905, 7–12; idem, "The Development of Aerology," *QJRMS* 53 (1927), 73–80.

progress in flight made the atmosphere above the ground increasingly a focus for investigation. Although use of manned balloons to obtain data had increased during the 1880s and 1890s, especially in Germany, where Kaiser Wilhelm II generously supported such efforts, the invention of the box kite in 1892 and the development of self-recording instruments had opened up additional possibilities for exploration.[52] By 1900 the kite had assumed a major role as an aerological instrument.[53] By this time the exploration of the upper air had also begun developing its own problems, methods, and social organization distinct from the broader profession of meteorology. In 1896, under the leadership of Hugo Hergesell in Strassburg, the International Commission for Scientific Aeronautics was established within the International Meteorological Committee. Enthusiasts such as A. Lawrence Rotch and L. P. Teisserenc de Bort used private funds to erect aerological observatories for exploring the upper air with kites and balloons. These centers at Blue Hill (Boston) and Trappes (Paris), along with those established with state funds by Richard Assmann, Hergesell, and Wladimir Köppen near Berlin, Strassburg, and Hamburg, spurred aerology's initial growth.

Although some traditional meteorologists took an active interest, most of the leaders came from outside meteorology proper. They were generally serious amateurs with some scientific background. Aerological and flight enthusiasts communicated freely, for neither flying nor the study of the upper atmosphere was then professionally exclusive. Most aeronautics associations contained a mixture of wealthy patrons, sporting enthusiasts, military officers, and scientists. Assmann and his aerological observatory in Berlin, for example, worked with Prussian aeronautical ventures. Hergesell, in Strassburg, enjoyed a close relationship with the local military balloon corps and with Count Zeppelin.

Even this early in manned flight, aeronauts were expressing a need for greater information about the atmosphere. Weather maps based on surface conditions were little use to them; they wanted knowledge, preferably predictions, of conditions aloft. Flying enthusiasts took an interest in the "scientific"study of the upper air in part, too, to legitimate their claims that flight represented a major advance in material culture. Aeronautics books and journals inevitably contained chapters and articles on "scientific ballooning" or "aeronautics in the service of

52. 40 Jahre: Berliner Verein für Luftschiffahrt (Berlin, n.d. [1921/22]), pp. 10–17; Assmann and Berson, Wissenschaftliche Luftfahrten, W. E. Knowles Middleton, Invention of the Meteorological Instruments (Baltimore, 1969), pp. 292–98 and, more generally, chaps. 8–10.

53. Reinhard Süring, "Die Beziehungen zwischen Meteorologie und Luftschiffahrt," Ill. Aër. Mitt. 2 (April 1900), 49.

science." Meteorologists tended to agree that observations from the upper atmosphere could be useful. After failing to find general laws for forecasting from weather maps based on surface data, they could dream of finding them in upper-air observations.

To develop as a vital scientific specialty, aerology required elaboration of its technological and organizational bases. Much of its data came from sporadic experiments, but systematic exploration and study of the upper air was not yet possible. Despite their promise as important aerological instruments, kites actually began to become a nuisance. A series of large box kites lifting instruments on three thousand meters of piano wire had an unpleasant tendency to fall on the ever-growing electric, telegraph, and telephone lines. Kites were not reliable, especially at inland stations, for use in regularly scheduled cooperative experiments nor in daily weather forecasting. Manned balloons were too expensive and cumbersome for frequent use. To get around these problems aerologists started experimenting with unmanned balloon systems for carrying self-registering instruments.

At the 1902 meeting of the International Commission for Scientific Aeronautics in Berlin, a breakthrough in the development of unmanned instrument-carrying balloons was announced.[54] Although Teisserenc de Bort and others had made huge paper "sounding balloons," these were expensive, clumsy, and not scientifically reliable.[55] Working with the Continental Rubber Company in Hanover, Assmann developed relatively inexpensive, reliable sounding balloons that ascended at fixed velocity. These could make possible frequent, regular upper-air observations at prearranged times from many locations. The conferees passed a resolution to publish rapidly the data collected during such internationally organized ascents. A poem was composed at the meeting in honor of the new balloon and of Assmann.[56] These rubber sounding balloons did in fact almost immediately confirm Teisserenc de Bort's unexpected, dramatic, and controversial discovery of a stratum in the upper atmosphere where the temperature stopped decreasing and seemingly began to increase with height: the stratosphere.

Optimism and expectation pervaded the meeting. Kaiser Wilhelm II hosted the congress and promised financial support for the commission's activities, including all the costs of data publication. The

---

54. A. L. Rotch, "The International Aeronautical Congress at Berlin," *MWR* 30 (July 1902), 356–62. The following discussion draws upon this article and the above-mentioned literature on aerology.

55. Middleton, *Meteorological Instruments*, pp. 301–6; Berson, "Aerologische Forschung," pp. 559–64; Linke, *Luftschiffahrt*, pp. 282–89.

56. "Das Lied vom Gummiballon," quoted in Linke, *Luftschiffahrt*, p. 288.

delegates, along with representatives from the government and military, were elated: the age of flight had begun; a new phase in the evolution of European civilization was underway. Pictures of Count Zeppelin's first giant airship, of various smaller, nonrigid airships including that of Alberto Santos-Dumont flying around the Eiffel Tower, and of military balloon corps were becoming commonplace in the popular press. Even at this early date, predictions of a forthcoming revolution in warfare and commerce captured the public imagination. Aerology seemed certain to grow in importance.

Bjerknes's Vision: Rational Predictions of Weather

Advances in aerology prompted Bjerknes to expand the scope of his engagement with atmospheric science. Traveling through Germany in 1902 to supervise the publication of the second volume of his work on hydrodynamic action at a distance, Bjerknes again visited colleagues. He arrived in Berlin shortly after the meeting of the international commission and received from Assmann and Köppen firsthand reports on the breakthroughs and on expectations for the future. Although international aerological programs and aerological instruments still required considerable elaboration before the upper air could conceivably be integrated into a scientific study of the atmosphere, Bjerknes now shared the aerologists' confidence that the time for this step was rapidly approaching.[57] With these advances in mind, Bjerknes formulated a project that, on the basis of the eventual attainment of sufficient upper-air data, aimed at establishing an exact mechanical physics of the atmosphere which could provide "rational" predictions of the weather.

Virtually from the start of his contact with meteorology, Bjerknes was intrigued with the thought of predicting atmospheric changes; he also understood from the start that all his notions required access to aerological data. He noted in 1898 that if the circulation theorem could account for the energy of cyclones, then with the help of kites it might be possible to determine whether a particular cyclone was still growing in strength or beginning to lose its source of energy.[58] At that time, meteorologists had no sure scientific means of predicting a cyclone's growth and decay. Bjerknes's hopes stemmed as much from an epistemological ideal as from any philanthropic desire to forecast weather: he considered these applications of his theorem just part of his broader goal of furthering mechanical physics based on con-

---

57. V. Bjerknes to H. Bjerknes, 18 June 1902, BFC; *DN*, 16 February 1903.
58. Bjerknes, "Dynamische Princip der Zirkulationsbewegungen," p. 156.

tiguous-action forces. Naturally he envisioned all these studies in terms of the book he took to be basic for all future theoretical physics: Hertz's *Principles of Mechanics*. Hertz had proclaimed that the highest ideal of mechanics is to be able to precalculate future events: "The most direct, and in a sense the most important, problem which our conscious knowledge of nature should enable us to solve is the antici- pation of future events, so that we may arrange our present affairs in accordance with such anticipation."[59] Bjerknes suggested in 1901 that meteorology, like mechanics, held prognosis as its highest ideal.

By identifying the prognostic problem in meteorology as one and the same with that in mechanical physics, Bjerknes began to conceive the problem of weather forecasting in terms of calculating changes in the atmosphere over time with the help of the circulation theorem.[60] This still vaguely formulated plan rested on the fact that the theorem contains time differentials and thus in principle is prognostic. Al- though at this point he agreed to assist Sandström in writing a book on methods for applying the circulation theorem, he had no further plans.[61] Even the preparation of a systematic treatise for using the theorem was hampered by the lack of suitable data for illustration. Indeed had Bjerknes been willing around 1900 to abandon his work on analogies, he could have done little more than sporadic applica- tions of the theorem. Use of his physics for a major endeavor in meteorology would require an array of synchronized upper-air obser- vations that was then unobtainable and not even realistically conceiv- able. The situation in oceanography for obtaining sufficient data and for regular use of the methods was even less satisfactory.

After his trip to Germany in 1902, Bjerknes knew that this situation was very much in flux; technological and organizational change was underway. He was prompted to consider a broadening of his concep- tion of applying physics to the atmosphere by a series of unusually severe storms in Sweden late in 1902 and again in 1903. The havoc enabled Ekholm to launch a campaign for a special storm-warning system for the Swedish coastal regions. Ekholm claimed that upper-

59. Heinrich Hertz, *The Principles of Mechanics, Presented in a New Form*, trans. D. E. Jones and J. T. Walley (1899; rpt. New York, 1956), p. 1. Bjerknes often quoted or paraphrased Hertz in discussions of prediction; see, for example V. Bjerknes, *Fields of Force* (New York, 1906), p. 154; idem, "Veirforudsigelse og muligheden for at forbedre dem," *Aftenposten*, 9 January 1904; idem, "Om veirforutsigelse som fysisk problem," *Naturen* 47 (1923), 43.

60. Bjerknes, "Cirkulation relativ zu der Erde," p. 102.

61. Nansen to V. Bjerknes, 12 March 1901, VBP, comments on Bjerknes's initial thoughts on a possible book of methods; V. Bjerknes to Ekman, 15 March, 26 October 1903, VWE.

air measurements, obtained by kite or balloon, could provide empirical clues of approaching storms otherwise undetectable until they had all but arrived.[62] Bjerknes saw this plan as typical of weather forecasting: it relied on methods based on empirical signs or statistical regularities without incorporating physical theory. Later in the year Bjerknes presented in a lecture at the Stockholm Physics Society his own vision for using upper-air observations in a forecasting scheme, one based on physics. The lecture, "A Rational Method for Weather Prediction," served as a basis for a series of popular and scientific articles in which Bjerknes explained his thoughts and aims for a major research program.[63]

Bjerknes turned to the possibility of transforming his efforts to comprehend atmospheric motions with the circulation theorem into a project for establishing a mechanical physics of the atmosphere. Assuming that observations from the upper atmosphere and over the oceans—to be obtained from ships with the aid of wireless telegraphy—would soon become regular features of the international exchange of weather data, Bjerknes asked whether forecasters would make good use of these additional observations. Although knowledge of the physics of the atmosphere had increased over the past half century, there had been little or no impact on forecasting methods. Bjerknes wanted to incorporate as much physics as possible into forecasting, which ought, he thought, to permit a mathematical solution to the forecasting problem. Two conditions were necessary for creating such a physics: knowledge of the state of the atmosphere at a given time and knowledge of the physical laws by which one atmospheric state develops out of another. In principle, the hydrodynamic equations of motion, the equation of state, and the laws of thermodynamics provided seven equations that could be used to calculate changes of the atmosphere's state as defined by seven atmospheric variables: temperature, humidity, pressure, wind velocity (three scalar quantities), and density/specific volume. The circulation theorem would be relegated to an auxiliary role in calculating relations among variables and physical laws. Additional possible influences such as

---

62. *DN*, 16 February 1903.

63. Bjerknes's lecture received special notice in the newspapers: "Rationell metod för väderliksförutsägelser: Föredrag af professor V. Bjerknes," *SD* and *DN*, 26 October 1903. The plan was more fully explained in V. Bjerknes, "Das Problem der Wettervorhersage, betrachtet vom Standpunkte der Mechanik und der Physik," *MZ* 21 (1904), 1–7, and idem, "Veirforudsigelse og muligheden for at forbedre dem," *Aftenposten*, 7–9 January 1904. Honoria Bjerknes related the family's experiences during the "terrible storms" of summer 1903 while they lived along the Swedish coast in a letter to Bjerknes's mother, Aletta, 1 September 1903, BFC.

irregular solar effects, electromagnetic fields, and heat radiation could be included later; Bjerknes considered these to be rather secondary.

Direct numerical calculation, of course, would not be practical; instead Bjerknes suggested creating graphic methods informed by physical laws. Using upper-air observations, one could construct such graphs for various levels above the surface. Consciously thinking of a medical analogy, he called for a complete "diagnosis" of the atmosphere's three-dimensional state at a given time, which would be used to attempt a "prognosis" of the change in the state of the atmosphere. Bjerknes appreciated the tremendous difficulties entailed in pursuing this vision of an exact physics of the atmosphere. He believed, nonetheless, that at least for certain variables it should soon be possible to produce graphic methods practical for daily weather forecasting.

The fate of "the problem of weather forecasting, considered from the point of view of mechanics and physics," as Bjerknes called his idea in one article, was inextricably linked with the availability of appropriate upper-air data, so Bjerknes joined the Swedish Aeronautical Society.[64] He, along with Arrhenius, Ekholm, and Sandström, endeavored to have the society participate in the international commission's efforts to send up across Europe manned and unmanned balloons as well as kites on the first Thursday of every month. Bjerknes addressed the Swedish Aeronautics Society on the inseparability of his proposed project for establishing rational weather forecasting and the use of aeronautics for obtaining upper-air observations.[65]

Bjerknes observed that by being restricted chiefly to observations from near the surface, meteorologists had not progressed much beyond "the old weather seers," for both had to rely on external signs that, with some degree of reliability, herald coming weather changes. The laws by which atmospheric conditions develop and change had consequently largely evaded meteorologists. With aerological observations available from but a few locations, only sporadic insights into the atmosphere's nature were possible. But change was afoot. Plans for international cooperation to obtain synchronized upper-air data marked "a decisive step . . . toward laying a fully rational foundation for meteorological research," a first step along "the only path" to an

64. Membership lists in the protocol for Swedish Aeronautical Society, cat. no. KSAK 3705, Technical Museum, Stockholm, shows that Bjerknes joined in 1904; Sandström and Arrhenius had already been members by 1903; Ekholm helped found the society in 1900.

65. V. Bjerknes, "Luftfärder i vetenskapens tjänst," *SD*, 26 March 1905. Much of the argument repeats Bjerknes's earlier statements in "Veirforudsigelse," 9 January 1904.

exact science of the atmosphere. Bjerknes had no doubts: "The problem of reliable weather predictions can be solved and shall be solved when the necessary material for its solution will be available in the form of a complete diagnosis of the atmosphere's actual [three-dimensional] state."[66]

Bjerknes understood that his idea of an exact physics of the atmosphere was, to say the least, grandiose. After his October 1903 lecture at the Physics Society, he wrote to Ekman that he had become occupied with the "thought of forecasting the weather mathematically,[67] but he did not immediately transfer his allegiance and energies to this project. He formulated the idea of an atmospheric physics; he regarded the idea; he weighed the possibilities and professional consequences of pursuing the idea. At the same time he increased his efforts to devise a common set of equations for electromagnetic and hydrodynamic fields of force. When during summer 1904 he was confronted with the latest advances by physicists toward replacing the mechanical foundations of physics, he naturally gave yet further thought to this project that could provide an alternative direction for his career. On the same late summer day in 1904 when he wrote in despair to Lorentz on the state of physics, he also wrote a long letter to Nansen:

> Since we last talked I have had a chance for the first time to gather my thoughts on the problems of meteorology and hydrography. And this has lead to a complete transformation of my view. I could no longer withdraw from the answer to that question: what do I really want [to do]? I found the answer could be only one: I want to solve the problem of predicting the future states of the atmosphere and ocean. I had previously closed my eyes to the fact that this actually was my goal, I must confess, partially for fear of the problem's enormity and of wanting too much.[68]

It would appear that while despairing over the fate of his mechanical physics, Bjerknes decided to experience how it might feel to commit himself to this project which loomed in front of him but to which he was not as yet ready devote his future.[69]

66. Ibid.; see also "Mitteilungen aus Schweden," *Ill. Aër. Mitt.* 7 (1905), 62, on Bjerknes's plans seen from the aeronaut's perspective.

67. V. Bjerknes to Ekman, 26 October 1903, VWE.

68. V. Bjerknes to Nansen, 4 September 1904, FNP.

69. V. Bjerknes to Ekman, 10 February, 18 March 1905, VWE; Stockholm Physics Society, "Protokoll"; V. Bjerknes to Lenard, 20 August 1905, PLP; V. Bjerknes to H. Bjerknes, 13 February, 11, 14 March 1904, BFC, all point to his continued preoccupation with the hydrodynamic analogies.

From summer 1903 until summer 1905, Bjerknes experienced frequent depression and insomnia. His frustrations in the world of physics were only part of a series of personal misfortunes that help explain his seeming vacillation and, as he remarked, uncertainty of purpose. Starting in spring 1903 his father died suddenly; his oldest son, Karl, who suffered from a number of congenital disorders, required a series of operations; his younger sons had trouble with their nerves because of Karl's constant problems; Honoria miscarried; and his father-in-law died. Bjerknes himself, with his relative poverty, felt a social outcast among his wealthier Stockholm colleagues. He had few friends. Personal and professional antagonisms toward him were growing in Sweden; although he claimed indifference, he apparently was deeply hurt and bitter at not being elected to the Nobel Committee for Physics. No wonder he complained of not being able to work at the same time that he was possessed by a great desire to throw himself into work.[70] He hoped with all his heart to succeed with his hydrodynamic actions at a distance, but he understood that the "*Zeitgeist*," and "energeticism and other mystical directions in physics," were against him.[71] He not only sought professional prestige for himself; he still felt compelled to vindicate his father's many years of ceaseless work that had exerted virtually no influence in physics. Clearly Bjerknes needed and wanted to break out of the situation in which he found himself, and the goal of an exact physics of the atmosphere could provide the opportunity. Events helped convince him that he could take the chance.

First to make progress on such a complicated project, Bjerknes would need help. Unable to manage the arduous calculations, he would have to depend upon assistants. Without Sandström or a similarly trained full-time employee, the project could not progress. Even with a generous grant from the Lars Hiertas Memorial Fund, Bjerknes could not offer a full salary to Sandström, whose expertise was also sought by Pettersson and various Swedish oceanographic and meteorological institutions.[72] Help came unexpectedly when, after presenting his lectures on hydrodynamic action at a distance at Columbia University in December 1905, he traveled to Washington, D.C., where at Abbe's request Bjerknes spoke on his vision for an

70. V. Bjerknes to Ekman, 20 October 1903, VWE; V. Bjerknes to A. Bjerknes, 20 December 1903, BFC; V. Bjerknes to Lenard, 11 May 1903, 4 April 1904, PLP.

71. V. Bjerknes to Lenard, 9 February 1906.

72. V. Bjerknes to Ekman, 5 April 1904, 10 February 1905, VWE; V. Bjerknes to Bjørn Helland-Hansen, 22 April 1904, BHH; V. Bjerknes to Nansen, 2 September 1904; Wilhelm Odelberg, *"En Fond som aldrig kan åldras": Stiftlesen Lars Hiertas Minne. En återblick, 1878–1978* (Stockholm, 1981), p. 157.

exact physics of the atmosphere.[73] With his polished and stimulating lecture, Bjerknes greatly impressed the audience. One listener, whom Bjerknes met through Abbe, suggested the possibility of financial assistance for the project. He was R. S. Woodward, president of the recently established Carnegie Institute of Washington and a mathematical physicist. At dinner the two found much to discuss and quickly developed a mutual liking and respect. Woodward invited Bjerknes to apply for a minor grant to support an assistant or two. Early in 1906, upon receiving an initial positive response from Washington, Bjerknes employed Sandström as a full-time assistant.

Second, the "conquest of the air" became a reality with the success of heavier-than-air flying machines. Aerology's vigor and social prestige were underscored when Kaiser Wilhelm II, with much publicity, presided over the opening in 1905 of the new Royal Prussian Aerological Observatory in Lindenberg, which he supported generously. At the same time Assmann and Hergesell founded a new journal through which aerology could express itself: *Beiträge zur Physik der freien Atmosphäre.* Bjerknes, seeing that this romantic, grandiose step for "civilization" could develop rapidly and successfully only if freed from the hazards of the weather, and claiming that meteorology and aeronautics "will always come to have a relationship of mutual dependency," foresaw that "the more aerial navigation develops to play a role for mankind—in war or peace—the greater will be the demand to know at any time the state and motion of the atmosphere so that the aerial voyage can be planned on the basis of this knowledge. And on the other hand meteorology, for its development is completely dependent on aeronautics in this word's broadest meaning. It alone can provide the observations that will allow us to study completely the atmosphere's laws."[74]

Bjerknes could finally consider devoting himself to a physics of the atmosphere because such a science would necessarily have great scientific, social, and cultural importance as aviation continued to develop. Equally decisive for him, the success and prestige of a new atmospheric science based on mechanical physics could also help legitimate retaining a mechanical foundation for physics.[75] His goal, after all, was

73. Discussion based on the following letters: V. Bjerknes to H. Bjerknes, 30 November, 3, 7, 17, 25 December 1905, BFC; Abbe to V. Bjerknes, 24 November 1905, VBP. Apparently A. L. Day asked Bjerknes during the summer to come to Washington from New York, and Abbe asked Bjerknes to speak on meteorology in addition to physics.

74. V. Bjerknes, "Om den videnskapelige luftseilas," *Samtiden* 20 (1909), 402–3. The same sentiments are found in his "Luftfärder i vetenskapens tjänst" and "Veirforudsigelse," 9 January 1904.

75. Bjerknes, *Fields of Force,* pp. 129–30.

not merely to apply physics to select atmospheric phenomena; it was to redefine meteorology, the science of weather, as a branch of physics. The effort would be as much a professional strategy as a scientific one. Meteorology had no clear disciplinary profile; it was often taught and pursued by astronomers, geographers, and statisticians. Bjerknes would make weather a problem in mechanical physics, a problem to be abstracted and analyzed in terms of the statics, kinematics, dynamics, and thermodynamics of the fluid of air enveloping the earth. In considering the atmosphere to be a mechanical system, he assumed mechanical determinism to be valid. Early in 1906 Bjerknes announced his intention to devote himself to this task.[76] Without abandoning the ideals of his father and of Hertz, without changing his professional identity, Bjerknes intended to claim a new domain for mechanical physics, and he expected to apply some of the lessons he had learned: he declared that although up to now he had had "ignorance" working against him, his aim henceforth was "to force" meteorologists to use mechanical and physical principles.[77]

76. V. Bjerknes to H. A. Lorentz, 16 June 1906, HAL; V. Bjerknes to Lenard, 9 February 1906, PLP.
77. V. Bjerknes to Lenard, 9 February 1906, PLP.

PART II

TOWARD REALIZING AN
ATMOSPHERIC PHYSICS:
THE QUEST FOR
AUTHORITY AND
RESOURCES (1906–1917)

[3]

Bjerknes and the Aerologists:
The Campaign for Absolute Units

W HEN Bjerknes received support from the Carnegie Institution in 1906 and committed himself to establishing an atmospheric physics, he divided the project into four subproblems:
statics, kinematics, dynamics, and thermodynamics. Together the
methods developed in each of these proposed volumes would constitute his *Dynamic Meteorology and Hydrography* (the latter expression
being the term then prevalent in Scandinavia for physical-dynamic
oceanography). Bjerknes's goal was to alter the nature of meteorology, including aerology. Although he recognized the importance
of the methods for oceanography, he did not himself become involved with oceanographic problems.[1] He noted that considerable
sums were spent on meteorology (and oceanography) around the
world with little effect, and for lack of "any leading idea, or any
tendency to take into application the 'intrinsic' laws of atmospheric or
oceanic processes, which we fully know."[2] He intended his project to
remedy this problem by "bringing our theoretical knowledge [of
physics] into practical application":[3] he named the Carnegie project
"Preparation of a Work on the Application of the Methods of Hydro-

1. Bjerknes anticipated the creation of a dynamic hydrography, but because of the
relative lack of regular surveying expeditions the immediate need to influence the
developing science was not as pressing as in the rapidly growing aerology. Moreover,
the potential prestige, as well as the ability to play on the potential practical significance
of introducing rational methods, was greater at the time in aerology. Bjerknes also
recognized that the efforts of Ekman, Helland-Hansen, and Sandström were already
laying the foundation for future use of the methods in oceanography.
2. Bjerknes to R. S. Woodward, 7 February 1906, copy, Carnegie file, VBP.
3. Bjerknes, "Vid. selvbekj," October 1910, posthumous papers, 9a, VBP.

dynamics and Thermodynamics in Practical Meteorology and Hydrography." As initially conceived, the project's objective was not the development of new theory but the development of new methods and techniques for extending the existing theory of mechanical physics to a new range of phenomena: the atmosphere and the oceans. Of course the problem of application would require research: to develop appropriate methods; to decide how to use them; and to select among methods so as to balance the long-term goal of attaining precalculations of changes of atmospheric state with the short-term goal of getting results as soon as possible to make practical work more rational.

Bjerknes and his assistants could not carry out this project in isolation. When Bjerknes and Sandström had first agreed to work together on a publication of formal methods—a "Dynamische Methoden der Meteorologie und der Hydrographie"—Bjerknes assumed the book would attract enough attention that the further elaboration of the methods through practical application could be left to others.[4] He soon faced a dilemma: To convince aerologists and meteorologists of the value of these methods, he needed illustrative examples. But to obtain the data with which the methods could be illustrated, he had to convince the international aerological community to make and publish their observations in absolute units, that is, units derived directly from fundamental units of length, mass, and time and not based on arbitrary numerical definitions. The internal requirements of the project thus forced Bjerknes to seek international cooperation. Bjerknes and his assistants' efforts during the first years of the project show just how this need for standardized international data developed.

In *Statics*, the first volume of *Dynamic Meteorology and Hydrography*, Bjerknes and Sandström developed diagnostic methods for representing the fields of mass and pressure.[5] Much of this work consisted of preparatory calculations and tables needed for laying a foundation for the rest of the project. Rather than rely on vertical height as a coordinate, they defined a measurement of gravity potential: the *dynamic meter*, a measure of the amount of work required to lift a unit mass from sea level to a point in space against the force of gravity. Because the intensity of gravity decreases from the pole to the equator, a unit mass must be lifted higher at the equator than at the pole if the same amount of work is to be performed and thereby attain the

4. Bjerknes to Bjørn Helland-Hansen, 22 April 1904, 16 May 1906, BHH.

5. V. Bjerknes and collaborators, *Dynamic Meteorology and Hydrography*, Carnegie Institution of Washington Publication No. 88, pt. I, V. Bjerknes and J. W. Sandström, *Statics* (Washington, D.C., 1910).

same *level surface,* a surface everywhere perpendicular to a plumb line. Although values of dynamic meters at every location would be very close to values of height in meters, they would not be the same. *Gravity potential* thus consists of lengths measured along a plumb line and therefore offers the only natural coordinates (a natural definition of horizontal and vertical) and the only measurement applicable for a full dynamic analysis of atmospheric motions.

The need to use gravity potential rather than vertical height had become apparent when Bjerknes and Sandström used the hydrostatic equation and the equation of state (the ideal gas law) to depict fields of atmospheric mass and pressure. The hydrostatic equation provides a relation among pressure, density, and gravity potential, assuming all other forces to be negligible. In preparing tables of concise values of these variables, Sandström and Bjerknes had to work with aerological data measured in height; to use the hydrostatic equation, they then had to introduce the acceleration of gravity, which further complicated the tedious calculations. By replacing height with gravity potential they could remove the unnecessary fourth variable and facilitate the calculations. Obviously, if aerological data could be published in dynamic meters rather than meters, Bjerknes's project would benefit significantly.

In *Statics,* Bjerknes and Sandström also presented an absolute unit for atmospheric pressure. Noting that measurements both in inches and millimeters of mercury were arbitrary units of atmospheric pressure, they used a unit of absolute measurement, the millibar (mb). They then provided methods for representing the pressure field. One was *baric topography,* a system of mapping that indicates the topographic height measured in dynamic meters of each of several standard surfaces of pressure (from 1000 mb up to 300 mb). That is, for selected values of pressure, the points in the atmosphere (or ocean) having this pressure form an isobaric surface. Such maps provide, then, a topograph showing the variations of height of a particular surface above (or below) sea level. A second method was *isobaric charting,* which shows surfaces of pressure at selected standard values of gravity potential (from sea level to 10,000 dynamic meters). Using these results they produced charts for fields of mass giving the average density of the mass of air between level surfaces (the difference of pressure between these surfaces) and the average specific volume of the air in isobaric "sheets" (the thickness of these "sheets" in dynamic meters). Again, Sandström and Bjerknes were confronted with the need for rational data: international observations of pressure would have to be measured and published in millibars, or centibars, to be readily useful for such analyses.

Even in this preliminary "clearing the ground [*rydningsarbeide*]" for the real work to be done in *Kinematics* and *Dynamics,* the dependency of the project on international aerological ascents was clear.[6] Most of the basic methods found in *Statics* had been developed by 1904, but lack of even minimally suitable data with which to illustrate the methods contributed to delaying completion of a manuscript until 1908.[7] The need to obtain truly simultaneous observations for defining an atmospheric state at a specific time and to have these observations in rational units became even more acute once work on *Kinematics* had begun.

Kinematics, the study of pure motions without reference to the forces producing them, entailed analyses of velocity and specific momentum to describe the instantaneous state of motion.[8] Whereas statics entailed purely diagnostic analyses because the hydrostatic equation possesses no time variable, kinematics by definition implicitly was open to both diagnostic and prognostic methods. Bjerknes aimed at developing geometric representations of the wind, which would be vector fields of motion. By assuming that the atmosphere is a fluid medium, filling all space, and that every moving particle of this medium has an invariable mass, the geometric properties of the field of motion are therefore intrinsically related to changes in the field of mass. That is, a knowledge of the field of motion implicitly contains a knowledge of the future field of mass. Although a fundamentally prognostic relation exists between motion and mass, Bjerknes and Sandström first treated special cases of the equation of continuity of mass in which the time element drops out, so that they could create geometric diagnostic principles for representing the field of motion. Fundamental to *Kinematics* was the development by Bjerknes with his Carnegie assistants Sandström and later Olaf Devik and Theodore Hesselberg, of sophisticated graphic techniques with which to apply the relevant physical principles directly onto synoptic charts. This meteorological "graphic algebra" and "graphic differential and integral calculus" corresponded, according to Bjerknes, to graphic statics and dynamics in technical mechanics.[9]

6. Bjerknes, "Vid. selvbekj."

7. Bjerknes to Woodward, 7 February 1906, copy, VBP; Bjerknes to Helland-Hansen, 22 April 1904, 16 May 1906, BHH; W. Barnum (Carnegie Institution) to Bjerknes, 23 January 1908, VBP.

8. V. Bjerknes and collaborators, *Dynamic Meteorology and Hydrography*, pt. II, V. Bjerknes, T. Hesselberg, and O. Devik, *Kinematics* (Washington, D.C., 1911).

9. Bjerknes expressed considerable enthusiasm over these new techniques; see "Vid. selvbekj."; Bjerknes to Svante Arrhenius, 5 November 1910, SAP; Bjerknes to Woodward, 7 January 1911, copy, VBP.

These innovative methods made it possible to represent wind observations, taken at individual points, as a field of continuous motion on two-dimensional planes. Special techniques had to be developed for analyzing vertical motions, which although small in comparison to horizontal motions, are responsible ultimately for the formation of clouds and precipitation. Because vertical motions could not be measured directly, Bjerknes and his assistants developed indirect methods for their analysis based on horizontal motions and the equation of continuity. By devising this representational form for the field of atmospheric motion using instantaneous lines of flow and lines of equal speed, they could map the field of motion at a given instant in time. Such lines of flow on a horizontal plane generally revealed points and lines where the air either converged or diverged. These, in turn, were associated with instantaneous upward and downward motions. Successive horizontal layers of the atmosphere could be mapped in this way, the three-dimensional atmosphere being defined in practice as the sum of the individual layers. Patterns in the lines of flow could be associated with idealized atmospheric motions such as a wave motion, convergence in a cyclone, and divergence in an anticyclone. But although they could construct these geometric patterns related to such motions, they needed actual illustrative examples both to explore the physical significance of the various kinematic patterns further and to demonstrate the methods' usefulness in practice. Once they turned to *Dynamics*—and with that, to actual attempts at calculating atmospheric changes—access to suitable data would only become more acute.

Dependency on International Aerology

Bjerknes was by 1909 clearly worried. He was aware that aerology, still in its formative years, had not as yet been molded in any particular scientific direction; its choice of problems and methods were still open to negotiation. He hoped to influence these choices before a consensus unfavorable to his project could form.[10] Indeed he discovered quickly that unless he could influence the international aerological community, he would never attain his professional goals. He had at first thought that if *Statics* were published early enough in aerology's development, its methods would exert an influence on the initial organization and publication of aerological observations,[11]

10. Bjerknes to Woodward, draft of letter, autumn 1908, and 13 May 1909, copy, VBP.
11. Ibid., 8 August 1910, copy, VBP.

"and we should then have been able to solve a series of important problems where we can now only give general indications as to their possible solution."[12]

Large amounts of money were being spent on observations of little "scientific" value as defined by quantitative, dynamic goals. Aerological observations were being made from a spectrum of institutions ranging from well-financed special observatories for "aeronautical meteorology" to military installations and ballooning clubs. National differences in instruments used and in methods for correcting instrumental errors created problems in analyzing European data: "When we draw our charts for the higher strata, a cyclone always appears over Strassburg."[13] The international organization showed negligible concern for strict simultaneity in coordinating ascents; publications of the aerological data provided no means of ascertaining the times when observations were taken. Bjerknes complained that "In as much as my work is dependent upon such observations it cannot bring more out of them than their quality admits."[14] Unlike Bjerknes, aerologists had only qualitative uses for data, such as ascertaining the existence and mapping the extent of the thermal-inversion layer in the upper atmosphere (stratosphere). Those physicists who occasionally took up atmospheric dynamics worked primarily with highly idealized situations and therefore were not so dependent on the availability of data. Bjerknes wanted to work with actual atmospheric states defined at specific times.

Bjerknes resolutely believed that if he could show the international aerological community how its data could be used in a rigorous scientific manner, it would agree to change to rational units and methods: "The organizators [sic] evidently have had no clear plan for the theoretical use of the observations produced. It has always been my intention to try to get an influence upon the organization of these observations as soon as the first part of my book had appeared."[15] It was bad enough that *Statics* had been delayed by the Carnegie Institution's publisher, in the meantime rapid advances were being made in aeronautics and aerology. Bjerknes was uneasy: "What makes the state of things still worse is the dependency of systematic meteorological and hydrographic observations on international agreements. If such agreements may be made and if they be made in the present state of confusion, the confusion can be made permanent, and the result can easily be a stagnation of these branches for many

12. Ibid., 1908, autumn draft.
13. Ibid.
14. Ibid., 8 August 1910.
15. Ibid., 13 May 1909.

years."[16] And with that, his scientific and professional hopes might once again be dashed, as he would not be able to carry through his project, nor would any of his methods be of interest or of use to others. He resolved to travel around Europe and try convincing aerologists of the need to adopt simultaneous observations measured in absolute units.

Aerology Emerges as a Vital Subdiscipline within Meteorology

The year 1909 marked a turning point in the development of aerology. When in that year Bjerknes addressed the sixth meeting of the International Commission for Scientific Aeronautics, he faced a confident assembly, proud of the recent progress in its field. The congress marked a change of emphasis in the relation between aerology and aeronautics. Previously aerology had been regarded chiefly as aeronautics's contribution to science or culture; now that flight had begun to be more commonplace, the role of aerology in making aerial navigation secure came increasingly into focus. Advances in aeronautics and optimistic plans for using flying machines militarily and commercially, including those for the establishment of a Zeppelin transport service (founded in November 1909 as the German Aerial Transportation Company—DELAG), posed challenges and opportunities for the aerologists.

Assmann, the head of the Lindenberg observatory and coeditor of *Beiträge zur Physik der freien Atmosphäre* raised this issue when he addressed the congress on "The Use of Aerological Observations in the Interest of Weather Prediction and Practical Aeronautics."[17] He indicated that rather than merely benefiting from aeronautics, aerology was in a position to be of service to aeronautics. Aerological institutions ought add to their scientific programs another one that would be "eminently practical": aerological observations for general weather forecasting and especially for specialized forecasting services for powered flight. He then described briefly a model service he had already organized. Whenever the Airship Study Association (Motorluftschiffstudien Gesellschaft) of Berlin was about to send up one of its Parseval non-rigid airships, aerological assistants sent up and tracked small pilot balloons from several stations around Berlin to measure wind conditions up to 2000 meters above the surface. They then telegraphed these observations as well as any indications of thun-

16. Ibid., 1908, autumn draft.
17. Richard Assmann, "Sur l'application des observatoires aérologiques à la prévision du temps et à la navigation aérienne," *Sixième réunion de la Commission internationale pour l'aérostation scientifique à Monaco du 31 mars au 6 avril 1909: Procès-verbaux des séances et mémoires* (Strasbourg, 1910), pp. 147–56.

derstorms or line squalls to the airfield. Assmann looked forward to the time when these observations could be even more valuable for aerial traffic once wireless telegraphy might allow direct transmission to airships. To serve this purpose aerological observations would have to be upgraded from sporadic organized "experiments" to a regular part of the regularly collected weather data. "In this manner the aerological observations will assume an important and essential share of the cultural work [*travail de la civilisation*] effected by aeronautics. And in accordance with this lofty and greatly significant goal, there can at present be but one request: Make the aerological observations permanent!"[18]

Assmann's goals were strongly supported by other delegates. Maj. H. Gross, chief of the German military air battalion and a designer of nonrigid airships, thanked Assmann for promising to help aeronautics and noted that Lindenberg's observations ought to be coordinated with the air battalion's maneuvers. The head of the Spanish airship program asserted that "without a continuous aerological service, a safe airship service is impossible," and meteorologist Peter Polis described the continuous aerological service for the Rhine provinces being planned by the Frankfort Physics Society in connection with the international aeronautical exposition that summer in Frankfort.[19] The discussion pointed to a converging of interests for establishing a permanent network of aerological stations both to assist practical aeronautics and to help explore the atmosphere's general circulation and structure.

Other matters were also reported and discussed at the congress. Aerology was clearly continuing to enjoy increasing prestige and growth. Just as aeronautical progress had become enmeshed in nationalistic endeavors for cultural and military superiority, so too did aerology find its place in this competition.[20] Aerological expeditions to tropical and subpolar regions won further insight into the structure of the trade and monsoon wind systems and of the stratosphere. Plans were also under way for using airships in arctic aerological expeditions. The widely publicized expeditions, the improvements in instrument design and construction, and the successful realization of

18. Ibid., p. 156.

19. *Sixième réunion de la Commission, pp. 35–37.*

20. See issues of *Beiträge zur Physik der freien Atmosphäre* during these years for descriptions and scientific results. The well-supported German aerological expedition to equatorial Africa in 1908 in which the newly built Ugandan railway brought the researchers to Lake Victoria Nyanza, among other similarly well-financed and publicized experiments, should be understood in the context of cultural and colonial hegemonic aspirations; see Lewis Pyenson, "Cultural Imperialism and the Exact Sciences: German expansion overseas, 1900–1930," *History of Science* 20 (1982), 1–43.

Beiträge zur Physik der freien Atmosphäre contributed to the enthusiasm of the scientists, military men, and aeronauts assembled at the congress in Monaco.

Bjerknes's Appeal for Reform

Eager to advocate reform in the aerological observations, Bjerknes had requested and received permission to attend the congress. Although excited about seeing the Riviera for the first time, he did not look forward to confronting the aerologists. When planning his trip he wrote to Arrhenius that he was going to "a meeting of the aeronautical meteorologists. I sometime must see that menagerie and speak a bit earnestly with them. It is surely a dreadful collection of dilettantes."[21] At the congress Bjerknes dissented in part from the general enthusiasm for establishing permanent aerological stations. He did not so much oppose this idea as assert that the commission must first institute changes in the manner by which aerological observations were made and published. To convince the commission of the importance of introducing simultaneous observations measured in absolute units, Bjerknes presented a talk, "On the Theoretical Application of Aerological Observations."[22] Using results from the as yet unpublished "Statics" and from the incomplete "Kinematics," Bjerknes tried to demonstrate the use of the new methods for analyzing aerological data. His program suggested what was perhaps the only systematic scheme for aerological data as an integral feature of a comprehensive atmospheric physics that would include practical, rational forecasting methods.

Unfortunately for Bjerknes, his methods and goals were alien to most of the delegates. Although he felt he had the audience in his grip, he did note that elderly meteorologists such as Hildebrandsson, who apparently did not know the difference between a scalar and a vector field, had a tendency to doze off during the talk.[23] Bjerknes's request that height be replaced with gravity potential made little sense to nonphysicists. Without copies of "Statics" to distribute to delegates who demanded to see the tables and graphic examples, he could only assure them that absolute units would enable the application of dynamic principles in aerology. Bjerknes's other request, for simultaneity of the international ascents, was easier to comprehend. For

21. Bjerknes to Arrhenius, 4 March 1909, SAP.

22. "De l'application théorique des observation aérologiques," *Sixième réunion de la Commission*, pp. 73–84. Bjerknes's use of "theoretical" in these contexts refers not so much to theory per se; rather, to the claim of being scientific as opposed to being purely empirical.

23. V. Bjerknes to Honoria Bjerknes, 2 April 1909, BFC.

obtaining a diagnosis of the atmosphere's three-dimensional state at a specific time over an area such as Europe or the United States, such an agreement was necessary. Still, some delegates nevertheless proposed using local time (i.e., all ascents to be made one hour before sunset local time) or questioned the need for absolute simultaneity.[24]

Fortunately for Bjerknes, he did not stand completely alone in the call for absolute units. Köppen, head of the Deutsche Seewarte in Hamburg and of its auxiliary aerological station, with whom Bjerknes had regularly conferred since they first met in 1899, also proposed introducing absolute units of atmospheric pressure. William Napier Shaw, head of the British Meteorological Office since 1900 and a former assistant to Maxwell, was on record in support of Köppen's proposal but could not attend the congress because of illness. Shaw believed that if for no other reason, meteorologists and aerologists ought to convert to absolute units to gain respectability in the world of science, where the use of such units forms a "freemasonry": by use of such units someone can be identified as "one of us or not".[25] Most important, the commission's president, Hergesell, recognized the potential seriousness of this issue for aerology's scientific profile, about which he was particularly concerned. He responded positively but cautiously to Bjerknes's lecture; he found it "very interesting."[26] By the end of the congress Hergesell engineered a general declaration on the potential benefit of using absolute units for certain purposes. Upon the commission's eventual approval, a proposal for use of these units would be sent for ratification to the Permanent Committee of directors of leading meteorological institutions in the International Meteorological Committee. Furthermore, the commission pledged to follow Bjerknes's proposals during the international ascent days planned for July 1909. The data thereby collected could then be used by Bjerknes to illustrate the advantages of his methods.[27]

Although pleased with this initial effort toward achieving the reforms, Bjerknes understood that he would still have to campaign to have these changes effected. At the very least, he needed to become more of an insider within aerology to be able to exert some influence. By understanding the problems aerologists faced, by partaking in aerological discussions and activities, and finally by attaining intellectual leadership within aerology, he might succeed. Before the close of the congress he took the first step: he promised to try to arrange for

24. *Sixième réunion de la Commission*, pp. 19, 34.

25. William Napier Shaw, "Pressure in Absolute Units," in idem, *The Air and Its Ways* (Cambridge, 1923), p. 9; orig. published in *MWR* (1914).

26. *Sixième réunion de la Commission*, p. 19.

27. Ibid., pp. 20, 41, 46–47.

Norway to join the expanding network of stations participating in the international ascent days. Bjerknes could not promise that Norway, which was once again his residence, would be able to participate, but he expressed a belief that if the government could be convinced of the "incontestable practical usefulness" of the aerological ascents for improving weather forecasting in general, there might be a chance of support. Bjerknes then was elected a member of the commission, a position he willingly accepted.[28] His campaign for absolute units had just begun.

The Campaign

Back in Norway, Bjerknes became involved with aerological and aeronautic activities; he began to become more of an aerologist himself. When discussion of forming a Norwegian ballooning society began in 1909, Bjerknes soon joined in planning the organization. At the inaugural meeting of the Norwegian Ballooning Society (Norsk luftseilasforening), large crowds listened as Bjerknes gave the opening lecture, "On Scientific Ballooning".[29] In it, he put the link between his project and the rapid growth of aeronautics sharply into focus.

Bjerknes began his talk by hailing the great step for civilization effected by "the conquest of the air"; he then added the further thought that we must hope the role of flying machines in war will not be as great as the fantasy writers have suggested. Regardless whether aeronautics's greatest impact will be for peaceful or for military purposes, he added, it would be too dangerous for any nation to fall far behind in the growing military competition, so the establishment of an aeronautics association "is a patriotic task of great significance." He then turned to the main subject of his lecture, the importance of aeronautics for meteorology.

In a comprehensive review of aerology's recent growth, and especially of the work of Assmann and Hergesell at Lindenberg and Strassburg, Bjerknes asserted that aerological observations and techniques for assisting flight, such as the sending up of pilot balloons to check upper-wind conditions along the planned path of an airship, can simultaneously serve science. Finally, he urged the association to collect funds to buy a balloon for manned flights. A balloon could be

---

28. Ibid., p. 34; Bjerknes to Woodward, 13 May 1909, VBP.

29. "Om den videnskabelige luftseilas," *Samtiden* 20 (1909), 401–12; Kåre Fasting, *Fra Kontraskjæret til Tokio: Norsk sivilflyging gjennom 50 år* (Oslo, 1959), pp. 18–24; Torleif Lindtveit and Gunnar Thoresen, *På vingene over Norge: Bilder fra flyhistoriens barndom* (Oslo, 1980), pp. 33–35.

of much more military and scientific value than an airplane because of its ability to maneuver in the vertical. He also urged the association to develop a pilot balloon program in connection with the international commission's ascent days: pilot balloons "are the only means that can provide us complete insight into atmospheric motions at all heights."[30] He especially desired the association to participate in the July experiment that was to be planned according to his own specifications. By doing so, they would at the same time obtain valuable information on upper-wind patterns over and around Norway's mountains, a necessity for the eventual full utilization of the manned balloon. Bjerknes's proposals were approved. Private funds for this very popular organization and project came quickly. In July, Bjerknes conducted the first Norwegian pilot balloon ascents, in cooperation with the International Commission.

Although Bjerknes was drawn to aeronautical activities in the hope of obtaining data for his project, and for displaying solidarity with European aerologists, he quite naturally joined the growing wave of excitement over flight. His wife, Honoria, had all she could do to prevent him from going up with the newly acquired balloon *Norge*.[31] When the first airplane flight in Norway was demonstrated in 1910 by the Swedish "flying baron" Carl Cederström, Bjerknes partook in planning the scientific, sporting, and social aspects of the occasion.[32] Still, the importance of joining the aeronautical activities stemmed from his need to understand the aerological requirements for flight. He could use the insights from these efforts for convincing the International Commission that his project could be of sufficient value to the commission's goals to justify its adopting the reforms in collecting and processing data that his project required.

America, Britain, and Germany

Bjerknes had few resources available for influencing various national meteorological and aerological communities other than the promise of his project's relevance for their own agendas. After the Monaco congress, he began his campaign for change almost immediately. First, he typed a nine-page letter to President Woodward of the Carnegie Institution.[33] In it he reviewed his efforts toward a kinematic diagnosis and prognosis of the wind field and then noted that the observations obtained from the International Commission had

30. "Om den videnskabelige luftseilas," p. 410.

31. H. Bjerknes to Maja Arrhenius, 27 March 1910, SAP.

32. Fasting, *Fra Kontraskjæret til Tokio*, pp. 67–69, 77; Bjerknes's assistant, Hesselberg, also participated enthusiastically in these aerological and aeronautical events.

33. Bjerknes to Woodward, 13 May 1909, VBP.

been "very bad" because of poor organization and lack of theoretical goals. He explained that until he had produced the volumes illustrating the methods and their application, he would have difficulty in convincing others to institute change; he wanted a set of good data, possibly from the United States, for use in *Kinematics*. American observations, because they covered such a large area under the leadership of one central weather bureau, could, if properly organized, offer such a set of data. Through Woodward's influence on the U.S. Weather Bureau, and possibly through supplemental aid from the Carnegie Institution, an experiment of limited duration could perhaps be instituted. Bjerknes pointed out that the recent advances in technique using rubber pilot balloons and theodolites allowed measurements of wind velocity aloft to be more accurate than those obtained at the surface with traditional anemometers. Bjerknes could not say in advance how many observation stations would be necessary to fulfill an approximation of a "continuous" depiction of air flow; he ventured to assume that eighty stations east of the Rocky Mountains should be sufficient. Similarly, if observations were made simultaneously three times daily during the experiment, these should enable him to apply his diagnostic and prognostic methods. Bjerknes justified the expense by noting the potential value for aeronautics:

> The cost of these investigations could be insignificant in comparison to the sums spent at present in meteorology on the useless accumulation of observations. The demands of practical aeronautics and of dynamical meteorology are here perfectly identical, and thus a cooperation only to be recommended. The demand of practical aeronautics for a more complete insight in the laws of the atmospheric motions will soon be so strong that investigations of this kind *must* be taken up, and in my opinion it would be the best if men of science took the initiative at once, not leaving the initiative to "practical men", who will spend uselessly large sums of money before they find the rational way of carrying out the investigations.[34]

Second, Bjerknes left the 1909 congress convinced that to have reforms instituted, he needed personal contact with leading aerologists and meteorologists. Only face-to-face with individuals and small groups might he show the advantages of his methods over the older ones.[35] In 1910 he initiated trips to present his work in London and Berlin. Shaw arranged for Bjerknes to present a lecture, "Synop-

---

34. Ibid.; the use of "rational" again indicates the effort of the scientist to establish authority and legitimacy over the "practical men."

35. V. Bjerknes to H. Bjerknes, 29 December 1911, BFC; Bjerknes to Akademiske kollegiet, 21 June 1912, copy, BHH.

tical Representation of Atmospheric Motions,"[36] at the Royal Meteorological Society. This lecture, like those in Berlin, was not simply a
presentation of his research; he was in a sense selling his methods
and, with them, a rationale for adopting absolute units.

With the use of lantern slides, Bjerknes discussed why a dynamic
approach was conceivable and desirable. Although sufficient aerological data were not available, he did use American surface data from
1905, obtained through Woodward, to show the new methods for
representing the wind field (Fig. 2). After pointing out how various
lines and points of convergence and divergence in the wind field
correlated with various weather phenomena, he noted:

> I digress a little here to draw your attention to a feature of considerable
> interest, in what promises to be in the future a department of practical
> life: charts of this description may be of use for aerial navigation. A
> Zeppelin airship travelling 15 metres per second (34 miles per hour) in
> still air, having to go from Dodge City in Kansas to Chicago, would take
> 23 hours if it followed the straight line between these places, but if it
> followed more nearly the lines of flow [in the wind field] it would take
> only 21 hours 15 minutes, assuming that the state of motion persisted
> during the voyage. The charts show that the state of motion does in this
> case persist for a large part of the time necessary for the journey.[37]

Then after sketching an optimistic plan for a science of the weather,
he turned to various problems confronting its realization, including
the fundamental issue: lack of simultaneity and absolute units in both
surface meteorological and aerological observations. After pleading a
case for reform he concluded, "I have no doubt that the time will
come, and come soon, when all required observations from the higher
strata will be obtained. The development of aeronautics will make
these observations not only possible, but also necessary."[38]

Bjerknes was not alone in recognizing the value of appealing to
aeronautics's presumed meteorological requirements to effect
change. Just before he left Britain he received a note from a young
British meteorologist, Ernst Gold, who was enthusiastic about
Bjerknes's plans to introduce rigor into meteorology: "I hope you
have a good time in Berlin and will succeed in converting them to
rational methods. Do not forget to mention the Zeppelin airship!"[39]

36. "Synoptical Representations of Atmospheric Motions," *QJRMS* 36 (1910), 267–
86.
37. Ibid., p. 282.
38. Ibid., p. 286.
39. Gold to Bjerknes, 31 May 1910, VBP.

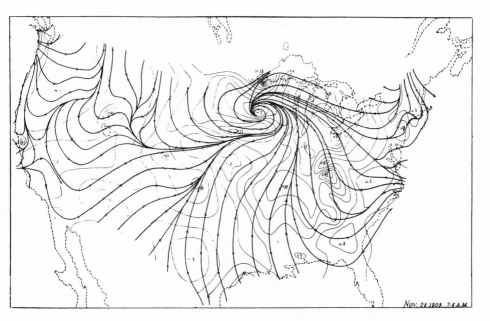

Fig. 2. Lines of flow and curves of intensity. The motion of air over the United States on 28 November 1905, showing lines of flow (heavy lines) and curves of intensity (thin lines). Lines of convergence enter the cyclone's center (upper Midwest), which is marked by a point of convergence in the lines of flow. On occasion Bjerknes used language from electromagnetism to describe the air motion: "It is seen how the 'lines of flux' run in towards the cyclonic centre" (Bjerknes to R. S. Woodward, 13 May 1909). From V. Bjerknes, "Synoptical Representation of Atmospheric Motions," *QJRMS* 36 (1910), 282.

Bjerknes did indeed have a good time in Berlin; and not simply because of the extravagantly expensive wine served during the reception after his lecture at the Berlin Society for Aeronautics.[40] His lectures in Berlin—the first one at the Berlin Society, the second at the Imperial Aero Club—were essentially the same as his London lecture. In addition he held private talks with meteorologists. Bjerknes convinced those present at the meetings that his method for diagnosing the wind field could at the very least play an "extraordinarily important role" in practical aeronautics;[41] he was elected corresponding member of the Berlin Society. Bjerknes still worried whether his pro-

---

40. V. Bjerknes to H. Bjerknes, 7 June 1910, BFC; V. Bjerknes, "Luftbewegung und Luftschiffahrt," *Deutscher Luftschiffverbandes Jahrbuch 1911*, pp. 3–14.

41. "Ueber Luftbewegung und Luftschiffahrt," *Vossiche Zeitung*, 10 June 1910.

posals would eventually be adopted, as he confessed to Arrhenius upon his return: "Absolutely amusing, it wasn't. An impression of hopeless dilettantism can be gotten everywhere"; although a hint of improvement could be seen in England under Shaw's influence.[42] Bjerknes felt he had to press on. His immediate goal was to have his proposals fully accepted at the 1912 meeting of the International Commission.

Third, Bjerknes attempted to circulate his work among the relevant practitioners, especially throughout Germany. When *Kinematics* was published in 1911, Bjerknes urged the Carnegie Institution to give priority to sending copies to meteorological and hydrographic institutions (rather than to university libraries) and to the leading men in these fields, for "only in this way will the work have an impact for changing observational methods."[43] More important, he urged the Carnegie Institution to allow a German translation and publication of the first two volumes of *Dynamic Meteorology and Hydrography*. Germany was the key to spreading his methods and, with them, his proposals; no other nation had so many aerological observation stations or provided so much money for oceanographic research. Once he received permission from the Carnegie Institution, Bjerknes contacted a publisher and described the role his proposed four-volume book could play in scientifically systematizing the growing aeronautical-meteorological, aerological, and oceanographic activities in Germany.[44] Translations of *Statics* and *Kinematics* were published in 1912.

Finally, Bjerknes focused his energies on the two leaders of German as well as international aerology, Assmann and Hergesell. Assmann opposed the introduction of absolute units. Hergesell was positively inclined but not committed. Bjerknes had to convince them that his methods could help resolve their problems. When Bjerknes learned of a new German weather warning system for aviators, established at Lindenberg in January 1911, he wrote Assmann asking for details of its aims. methods, and problems.[45] This so-called aeronautical weather bureau was to provide upper-wind conditions and warnings of thunderstorms and squalls to aviators. Although the system covered only the Prussian sphere of interest, Assmann hoped eventually to include the entire Reich.[46] Bjerknes wished Assmann

42. Bjerknes to Arrhenius, 5 November 1910, SAP.

43. Bjerknes to Woodward, 19 April 1911, copy, VBP.

44. Bjerknes to Friedrich Vieweg & Sohn, 3, 23 August 1911, copies, VBP.

45. Assmann to Bjerknes, 16 May 1911, VBP, discusses the contents of Bjerknes's letter, 6 May 1911.

46. "An Airman's Weather Bureau," *Scientific American*, July 1911, pp. 98–99.

success and offered assistance if he could be of help.[47] He arranged to visit Lindenberg later in the year. In advance of his visit, he sent one of his Carnegie assistants, Hesselberg, both to Lindenberg and to Hergesell's Strassburg observatory to become acquainted with the methods and problems of "practical aerology" at the two institutions.[48]

Toward the end of 1911, when Bjerknes himself visited the Lindenberg observatory, he was shocked to find an atmosphere of gloom and despair.[49] Although Assmann had about fifty assistants, "who live like cats and dogs," he could not get competent help. After having received lavish support for his observatory, Assmann feared the loss of funding and prestige for lack of impressive results. If Bjerknes could only help "get things going so that something is accomplished," Assmann would then "gladly support" Bjerknes's own efforts and project.[50] With his former opponent on absolute units "heavy on the hook," Bjerknes willingly exchanged a pledge to assist where possible, cautioning that he could not promise immediate success. It appeared that Bjerknes would "hardly get any more opposition from Assmann".[51]

Next he visited Hergesell in Strassburg. Impressed, and amused, by Hergesell's tireless travels and activities in support of the International Commission, Bjerknes sought to capture "Hin- und Her-gesell" for his own campaign. He had learned enough practical aerology to show Hergesell how rational methods and absolute units could save the Strassburg Observatory much labor. Hergesell, for his part, showed Bjerknes that the observatory's methods were not quite as hopeless as Bjerknes assumed. He agreed nevertheless to support Bjerknes's proposals fully and push for their acceptance at the next meeting. Hergesell urged Bjerknes to return to Strassburg before the commission's 1912 Vienna meeting so they could review and coordinate strategy. Bjerknes felt he had achieved his major goal for the trip and that the direct contact had been essential.[52]

Returning from this latest effort as a "missionary among the heathens," Bjerknes claimed that he had formulated the proposals for reform in as "innocent a form as possible" to have a chance of attaining "total victory."[53] Yet as the 1912 meeting approached, Bjerknes

47. Bjerknes to Assmann, 23 May 1911, draft, VBP.
48. Plans for Hesselberg's trip related in Assmann to Bjerknes, 29 January 1912, VBP; Hesselberg to Bjerknes, 12 March, 8 April 1912, JBP.
49. V. Bjerknes to H. Bjerknes, 22 December 1911; see also ibid., 26 April 1909; both in BFC.
50. Ibid., 22 December 1911.
51. Ibid.
52. Ibid., 29 December 1911.
53. Bjerknes to Arrhenius, 1 January 1912, SAP.

still had doubts: "The matter is not finished because on the commission sit an awful number of ignoramuses, balloonists, officers, etc. and nobody can judge in advance the result."[54] How would the older meteorologists react to his proposal to remove height as a primary independent variable and replace it with gravity potential? Of the sixty or seventy participants, perhaps only ten would even understand what gravity potential meant.[55] Hergesell agreed to allow Bjerknes to propose that in all the commission's publications, pressure should be expressed in rational units, and that gravity potential should be introduced as the fundamental variable instead of height.[56] Acceptance of rational units by the aerologists would almost certainly lead to universal acceptance in all meteorology, Bjerknes believed, because aerology was then the most vigorous part of meteorology and most likely to experience major development.[57]

The Immediate Results: Vienna 1912 and Rome 1913

Bjerknes's discussions with Hergesell, among others, proved helpful. At the opening of the Vienna meeting, which was attended by members of royalty, government, and the military, Hergesell reviewed the progress of aerology since the last meeting and the prospects for further advances.[58] He included Bjerknes's methods among the most promising developments, for they would allow full use of the international network of aerological stations and could provide detailed knowledge of the wind field at different levels of the atmosphere. While alluding to these practical benefits Hergesell also emphasized for the gathered dignitaries and aerologists that the work of the commission and of aerology ought first and foremost to aim at furthering *Wissenschaft*. Pursuing science for its own sake, was certainly necessary if aerology was to achieve academic respectability and legitimacy. Hergesell pointed out too that the recent deaths of aviators in weather-related accidents pointed strongly to the need for further study of atmospheric turbulence and squalls. Bjerknes's own presentations at the meeting followed the same rhetorical pattern.[59]

54. Bjerknes to Vilhelm Carlheim-Gyllensköld, 15 January 1912, KVA-SUB.

55. Bjerknes to Arrhenius, 1 January 1912, SAP.

56. Bjerknes to Woodward, 29 January 1912, copy, VBP; Bjerknes to Arrhenius, 8 May 1913, SAP.

57. Bjerknes to Arrhenius, 8 May 1913, SAP.

58. Hergesell's opening remarks, *Siebente Versamlung der Internationalen Kommission für Wissenschaftliche Luftschiffahrt in Wien 28. Mai bis 1. Juni 1912: Sitzungsberichte und Vorträge* (Vienna, 1912), pp. 19–29; his talk, "Einrichtung eines internationalen Netzes von Pilotballonstationen," ibid., pp. 135–36.

59. "Begründung seiner Vorschläge," *Siebente Versamlung der Internationalen Kommission*, pp. 73–77; "Ueber die synoptische Darstellung des Druckfeldes in der At-

In his proposal for absolute units he again pointed out the need to create a physics of the three-dimensional atmosphere. With the proper observations, he explained, it should be possible to replace meteorology's two-dimensional maps of isobars (lines of equal pressure) with aerological maps of three-dimensional surfaces in the pressure field. He spoke of creating a new geometry of the atmosphere.

The commission accepted his proposals, but with a condition: use of absolute units of pressure would have to wait until the Permanent Committee of the parent organization approved the change. The commission recommended too that Bjerknes's dynamic height, or gravity potential, be used in all its publications as of 1913 and that pressure should replace height as the independent variable in the published results of aerological ascents. To Bjerknes, the resolution implied that the commission had "in reality determined that the methods worked out in my book shall henceforth be used in international aerological work."[60]

But Bjerknes was too optimistic. Before the 1913 meeting of the International Meteorological Committee in Rome, the proposal to use absolute units of pressure came under attack in the prestigious *Meteorologische Zeitschrift*.[61] Although the head of the Austrian meteorological service, Wilhelm Trabert, wrote the attack, Bjerknes was sure that behind him was the grand old man of meteorology who represented the views of the old guard, Julius Hann.[62] Trabert argued that conversion to millibars would make comparison between old and new observations impossible. Moreover, the change would necessitate redefining the accepted value of standard atmospheric pressure, which in turn would mean changing all the calibration scales for instruments. Bjerknes, in responding to this "idiotic polemic," showed that Trabert misunderstood the meaning of converting to absolute units and that his fears were unfounded.[63] To help the cause of absolute units, Shaw arranged an invitation for Bjerknes to come to the meeting in Rome and lobby for change.

At the meeting, Bjerknes had to accept a partial victory. The committee did agree that all aerological observations should be published both in millibars and in the traditional millimeters. The proposal that

mosphäre," ibid., pp. 78–80; "Ueber eine Arbeit von Th. Hesselberg: Ueber die Luftbewegung im Cirrusniveau und die Fortpflanzung der barometrischen Minima," ibid., p. 146.

60. Bjerknes to Woodward, 25 June 1912, copy, VBP.

61. W. Trabert, "Millimeter oder Millibar?" *MZ* 29 (1912), 401.

62. Bjerknes to Arrhenius, 8 May 1913, SAP.

63. Bjerknes to Walfrid Ekman, 22 January 1913, VWE; Bjerknes, "Millimeter oder Millibar?" *MZ* 29 (1912), 576–78; idem, "Das C.G.S.-System und die Meteorologie, ibid. 30 (1913), 67–71.

dynamic meters replace meters of height, which had received little open criticism in Vienna in 1912, was now the cause of considerable controversy. Hergesell, in the name of the commission, withdrew the proposal until some future date.[64] Bjerknes was not as despondent as might have been expected. He knew that the American and British meteorological services were planning to implement absolute units in their publications of data. More important, he had already assumed a new position that not only would provide him with significantly greater authority within aerology and meteorology but would also offer him for the first time major resources with which to effect change. Bjerknes had become the director and professor at the world's first academic institution for training aerologists and for grounding practical aerology in a scientific foundation: the new Geophysical Institution of the University of Leipzig.

64. V. Bjerknes to H. Bjerknes, 9 March 1913, BFC; the proposals were not to be fully accepted until 1929.

[4]

Bjerknes in Leipzig: Resources
for Shaping a New Meteorology

THE significance for Bjerknes of receiving a call to set up a geophysics institute in Leipzig can best be appreciated by turning back to 1906 and examining the institutional constraints on his project. At roughly the same time in early 1906 that Bjerknes decided to turn to his atmospheric physics project, a group of Norwegian professors was attempting to establish a personal professorship for him at the Royal Frederik University in Christiania. As early as 1900, Nansen had headed a group of natural science professors in a bid to establish such a professorship for Bjerknes and thereby bring him home from Sweden. Although all authoritative bodies within the university had endorsed this early proposal, the Ministry of Church and Education (KUD) had refused.[1] In 1902 the death of C. M. Guldberg left open the ordinary professorship in applied mathematics, but Bjerknes confided in Nansen that he was not interested because the teaching responsibilities would curtail his research endeavors. Bjerknes still hoped too that Stockholm would soon become a major research center, and he would benefit from being there.

After Norway became independent from Sweden in 1905, leading professors in science and mathematics again proposed that Bjerknes should be called home.[2] National pride obviously was at stake ("We

1. Kristine Bonnevie to Honoria Bjerknes, 28 March 1900, BFC, describes the effort. Bonnevie, who became Norway's first woman professor in 1912, was Honoria's younger sister.

2. Materials related to Bjerknes's call to a personal professorship can be found in app. 7, 4 April 1906, and app. 8, 31 May 1906, in KUD/E, folder: "Universitetet, Budsjetter 1905/06–07/08."

can't afford not to have him here"),[3] but the sensitive political situation made it necessary to claim in public that "the Fatherland must have you [Bjerknes] back on purely scientific grounds."[4] The Academic Council, (Akademiske kollegiet) in its appeal to the ministry, emphasized that Bjerknes's work on electric waves was important for wireless telegraphy; that on hydrodynamics, for meteorology and oceanography. It asserted that his unique attempt at systematic analyses of the growing body of international observations obtained by balloons and kites, and of the observations obtained in north European oceanographic investigations, "promise a rich dividend for science and commerce." It added that "a land that has respect for itself generally tries to call home its leading sons who in other nations have achieved great fame."[5] Bjerknes received a personal professorship. When asked how it should be designated, Bjerknes requested that it be defined as a professorship in mechanics and mathematical physics, which should not be surprising in light of his goal of integrating the atmosphere and oceans into physics.[6] He did not, however, receive the conditions for steady progress toward this goal.

When, in September 1907, Bjerknes moved to Norway, he brought Sandström with him. Sandström had only a small salary based on Bjerknes's Carnegie grant and no guarantee of support beyond an initial five years. When Sandström received an offer in 1908 of a full-time job with the Swedish Hydrographic-Biologic Commission, he returned to Sweden. Unable to calculate tables or construct graphs himself, Bjerknes was almost helpless without Sandström. He soon found two able students, Olaf Devik and Theodore Hesselberg, and hired them as Carnegie assistants. But before they could be put to work, they first had to learn the project's special techniques and problems. Although they soon proved very capable, Bjerknes feared "great difficulties" should they also seek better paying or more permanent work.[7] State funds for permanent full-time assistants were virtually nonexistent in Norway, and the scarcity of students in natural science meant an uncertain pool from which to recruit replacements. Fearing the loss of Hesselberg, Bjerknes instead lost Devik in 1911.[8] Fortunately, another exceptionally bright student, Harald Ul-

3. Bonnevie to H. Bjerknes, 21 February 1906, BFC, quoting W. C. Brøgger.
4. Nordahl Wille to V. Bjerknes, 23 February 1906, VBP.
5. KUD/E, app. 8, 31 May 1906.
6. Wille to Bjerknes, 23 February 1906, VBP.
7. Bjerknes to R. S. Woodward, 8 August 1910, copy, VBP.
8. By 1910, Hesselberg was finishing his studies and planned to become engaged to be married; to hold him, Bjerknes felt obliged to seek an increase in funds from Carnegie (Bjerknes to Woodward, 8 August 1910, copy, VBP). Although Bjerknes

rik Sverdrup, was attracted to Bjerknes's project. Still Bjerknes found the situation unsatisfactory.

Even when he and his assistants could work uninterruptedly conditions were not conducive to rapid progress on the project. Having only a cellar room at the university, they were hampered by lack of space to assemble all the observations, tables, graphs, and maps.[9] Honoria assisted with the graphic work because of the lack of funds to hire additional help.[10] Because of the dwindling number of students, Bjerknes could no longer use his lectures to systematize and clarify his research work. Work on *Dynamics,* perhaps the most difficult and complex of the four volumes of his overall project, seemed threatened. Clearly, he had little chance to develop a school of disciples who could help spread the methods. He feared constantly that progress on his increasingly complex and demanding project, on which his scientific career now depended, could cease at almost any time.[11]

In Norway, Bjerknes gradually began to experience the same sense of powerlessness he had experienced in Stockholm as a theoretical physicist. Lack of direct access to journals and lack of adequate research facilities and sufficient assistants limited his ability even to influence the emerging aerological subdiscipline, let alone transform meteorology in general. Practical aerological activities were increasing dramatically in Germany, but regular direct contact was impossible because of the distance. Although in principle the acceptance of absolute units in aerology ought to make his methods more easily assimilable, Bjerknes recognized that in practice the conservatism and lack of rigorous scientific training of most aerologists virtually precluded the general acceptance of his atmospheric mechanics.[12] In summary, by 1911, Bjerknes found that the aerologists' and meteorologists' understanding of his work was limited, his ability to complete the remaining volumes of the project constrained, and the possibility of establishing a new generation of scientifically trained aerologists bleak. All this changed when Bjerknes was called to head a new geophysics institute in Leipzig.

---

received an increase of support from $1200 to $1800, from which salaries and research expenses had to be covered, Bjerknes could not hold Devik, who was recruited by Kristian Birkeland, as a privately financed assistant for his research on northern lights.

9. Fridtjof Nansen described Bjerknes's working conditions in a letter to W. C. Brøgger, 8 November 1909, *Fridtjof Nansen Brev,* ed. S. Kjærheim, vol. 3, *1906–1918* (Trondhjem, 1963), pp. 135–36.

10. H. Bjerknes to Maja Arrhenius, 31 August 1909, SAP.

11. V. Bjerknes to Woodward, 3 May 1907, 25 June 1912, copies, VBP; V. Bjerknes to Svante Arrhenius, 23 December 1907, 29 November 1908, SAP.

12. Bjerknes to Akademiske kollegiet, 21 June 1912, copy, HHP.

Origins of the Institute and Bjerknes's Appointment

The decisions to establish the Leipzig institute and to call a Norwegian, Bjerknes, to be its first director were related to developments in and further expectations for German aeronautics. From the start of "the conquest of the air" in the late nineteenth century, Saxony had been eclipsed by Prussia in this enterprise. To bring Saxony into the forefront of aeronautical culture, the newly founded Leipzig Society for Aeronautics (1909) embarked on several projects. Leipzig was to become a major airship center. Plans were drawn up for what eventually, in 1913, became the world's largest Zeppelin-airship hangar, part of the commercial airship service, DELAG.[13] In addition, in a report to the Saxony Ministry of Culture, the society stressed the importance of establishing at the university a professorship in "exact meteorology."[14] The chief advocate of the latter proposal was Otto Wiener, professor of experimental physics at Leipzig University and a founder of the aeronautical society.

Wiener believed in the importance of aeronautical progress as part of the further development of German culture. He was concerned that the significance of aeronautics was generally underestimated and troubled that aeronautics's dependency on weather and wind conditions had been largely ignored.[15] He urged the university to establish an institute for geophysics, with a focus on aerology.[16] It would provide theoretical underpinnings and direction to the ever-growing aerological and aeronautics-oriented meteorological activities in Germany. His dean's report to the Ministry of Culture asserted that aeronautical progress was as dependent upon dynamic meteorology as on technological improvement of aircraft structure. Establishing a professorship and an institute in this young field might contribute to solving some of the formidable problems confronting the normalization of flight.[17] Wiener had in mind a candidate for the position from the start: Bjerknes.

Wiener and Bjerknes shared a common interest in mechanical physics. Primarily an experimental physicist himself, Wiener pre-

13. Otto Wiener to Bjerknes, 27 September 1912, VBP; "Die Eröffnung des Luftschiffhafens in Leipzig," *DZL* 17 (1913), 344.
14. Wiener to Bjerknes, 21 October 1910, VBP, relates these events.
15. Ibid., 7 March 1911.
16. Ibid., 21 October 1910.
17. Dekan phil. Fakultät (A. Köster) to Königliche Ministerium des Kultus und öffentlichen Unterrichts, 23 May 1912, refers to contents of reports from the philosophical faculty written in January and June 1911, Manuscript Department, Karl Marx University (KMU) Library, Leipzig. I thank the Norwegian Research Council for Science and the Humanities for assistance in obtaining documents from the German Democratic Republic.

ferred Bjerknes's uncomplicated "classical" mathematical treatment of physics to the newer theoretical physics that relied on highly sophisticated mathematics. Wiener had, as I have noted, even considered Bjerknes as a possible candidate in 1902 for the chair in mathematical physics at Leipzig. He followed Bjerknes's gradual change of research priorities and approved Bjerknes's aim of extending mechanical physics to the atmosphere. After surveying the efforts in German meteorology, Wiener claimed that nobody during the past decade had approached the scientific quality of Bjerknes and Sandström's studies. Wiener was appalled by the marginal influence these studies had so far had.[18] Describing Bjerknes as the world's most outstanding researcher in dynamic meteorology, he claimed that, in the right position, Bjerknes could lead the way "to a new epoch in meteorology,"[19] a necessary precursor to the growth of aeronautics. The proposal to give Bjerknes the professorship claimed *Statics* had established a foundation for introducing absolute units into meteorology and that *Kinematics* provided new methods for graphic differentiation and integration that ought lead to an immediate revision of weather map analysis. The proposal gave emphasis to the significance for aeronautics of the Bjerknes methods of analyzing the wind field: several hours could be saved on a Zeppelin airship voyage if the course was defined by streamlines in the wind field rather than by the shortest geometric distance.[20] Bjerknes received the call.

He conceived his appointment as a means of acquiring resources not only to complete his project but to shape meteorology, including aerology, as a scientific discipline. This strategy is evident from his decision to accept the professorship, his understanding of the position, and his plans for the institute. As early as 1910 he knew Wiener was planning this professorship. In response to Wiener's preliminary inquiries, he noted his reluctance to leave his personal professorship in Norway but agreed to consider the position if offered. When Bjerknes traveled around Germany in December 1911 to gain support for absolute units, he stopped off in Leipzig for what he hoped would be a bit of relaxation. Instead, he found the normally easygoing, pleasant Wiener eagerly waiting to persuade him to make a commitment.

With the dismal situation at home, Bjerknes could no longer easily reject Wiener's offer. He wrote two letters to Honoria in which he considered for himself and for her the arguments for and against the Leipzig job. He admitted he would have liked to be able to say out-

18. Wiener to Bjerknes, 21 October 1910, VBP.
19. Wiener to Dekan phil. Fakultät (C. Chun), 14 June 1911, KMU.
20. Dekan to Ministerium des Kultus, 23 May 1912, KMU.

right no, but the family's "wretched" economic situation forced him to think more carefully. Upon further reflection, one point could not be avoided: "that at home [in Norway] I work in a void, but here [in Leipzig] I would get influence. According to Wiener's plan I shall reform all of Germany's meteorology and through that the entire world's."[21] He continued in his typical understated, humoristic manner, "And that could, after all, have something attractive about it." Turning to a consideration on which he repeatedly reflected, he asked his wife how they could best serve Norway: by remaining at home amid economic worries and uncertainties or by accomplishing something abroad? Wiener believed this "mission" could be completed within ten years; they could then return to Norway having established a new scientific meteorological discipline. Knowing her husband's dreams, frustrations, overexertions, and stubbornness, Honoria simply responded that if fate was to take them to Leipzig, so be it; she would follow him once again out into the world.[22]

Bjerknes could hardly refuse the Leipzig offer, for its two goals—to provide a scientific basis for practical aerological activities and to train people for the ever-growing aerological/aeronautical meteorological services—meant it could provide him the resources necessary to advance and spread his atmospheric physics, or his "dynamic meteorology," as he began calling the overall endeavor.[23] He informed colleagues that he would prefer not to leave Norway but that "this professorship is precisely in my line, aerology," and by being in Leipzig, "I shall easily come into contact with practical aerology, while the same thing will here remain very difficult."[24] Equally significant, by teaching his methods in Leipzig he could ensure the victory of absolute units and rational methods because his specially trained students would surely become the future leaders of aerology and meteorology.[25] He had begun to recognize clearly that "in itself the formal

21. V. Bjerknes to H. Bjerknes, 22 December 1911, BFC.

22. Ibid., 29 December 1911, repeats her reply, BFC. Bjerknes hoped that both the considerably higher salary and Leipzig's cultural offerings would make her amenable. V. Bjerknes to Fridtjof Nansen, 18 January 1902, FNP, offers another illustration of Bjerknes's including Norway's cultural development as one of his considerations in making career choices.

23. Bjerknes may well have sensed that professional physics was already narrowing its boundaries around a few specialties; his earlier hopes for a well-defined cosmic or terrestrial physics eventually claiming a significant place in the structure of the discipline were surely dampened. He interchanged the expressions "dynamic meteorology" and "exact physics of the atmosphere."

24. Bjerknes to Woodward, 25 June 1912, copy, VBP; see also Bjerknes to Napier Shaw, 1 July 1912, copy, VBP; Bjerknes to Arrhenius, 18 July 1912, SAP.

25. Bjerknes to Woodward, 6 May 1913, copy, VBP.

acceptance of a proposal means very little; it is the implementation that is crucial."[26]

To his long-time friend Arrhenius, Bjerknes was more direct in describing the attraction of the institute and professorship: at stake was control of the enormous resources available with the position. The institute, which would receive "considerable funds," could be organized according to his own wishes. He would be able to collect large amounts of aerological observations and then get the methods he and his assistants developed "applied in practice right away and all under my own control." After having "sailed mostly against the wind," Bjerknes felt no need to apologize for seizing this opportunity.[27] Because the Leipzig institute was to be the first major academic training and research center (as opposed to merely a professorship) specifically for atmospheric science, its director might well influence the development of aerology by educating a new generation of meteorologists.[28] Were he to turn down the position, someone else would receive it and with it the major influence. His choice was clear; he had to leave home.

Bjerknes's Leipzig School

Bjerknes arrived in Leipzig in January 1913. His inaugural lecture was a carefully worded treatise intended to place his institute in a Wilhelmian neohumanistic university tradition: classical allusions, a historical perspective, and for this obviously all too practical science, a vision of a far distant goal he scarcely could expect to reach in his own lifetime. In it he noted that the physics of the atmosphere, which he was attempting to establish, and meteorology treat the same subject; "however the two sciences must not be confounded with each other. The distinction is marked for the reason that physics ranks among the so called exact sciences, while one may be tempted to cite meteorology as an example of a radically inexact science. Meteorology becomes exact to the extent that it develops into a physics of the atmosphere." Where previously physicists treated highly idealized atmospheric situations, now, because of aerology, it becomes possible to apply the equations of exact physics to "the actual existing atmospheric conditions. It is for us to discover a method of practically utilizing the knowledge contained in the equations." While keeping the long-term goal of an exact physics of the atmosphere in sight—"steering by the stars"—he and his as-

26. Bjerknes to Akademiske kollegiet, 21 June 1912, copy, BHH.
27. Bjerknes to Arrhenius, 18 July 1912, SAP.
28. Bjerknes to Akademiske kollegiet, 21 June 1912, copy, BHH.

sistants would formulate and solve a series of intermediary preparatory problems leading toward rational prognoses of atmospheric conditions.[29] The immediate tasks in front of Bjerknes and his assistants were preparing the *Dynamics* and *Thermodynamics* volumes of the *Dynamic Meteorology and Hydrography* and beginning the mass production of aerologists trained in rational methods.[30] Starting with seven empty rooms, Bjerknes proceeded to build an institute that could fulfill his personal mission.

Even before getting fully organized, Bjerknes; his two Carnegie assistants, Hesselberg and Sverdrup; and his new institute assistant recommended by Hergesell, Robert Wenger, began work. Without furniture, they set about producing a publication series consisting of detailed analyses of aerological data using the methods set out in *Dynamic Meteorology and Hydrography*. These *Veröffentlichungen des Geophysikalischen Instituts der Universität Leipzig* included a series of charts depicting major atmospheric variables for each of several dynamic heights in the atmosphere for each day of the international aerological ascents. These charts were the first detailed, comprehensive aerological analyses ever produced; and their large, high-quality, two-color format helped establish them immediately as extraordinary resources for meteorology.[31]

Bjerknes was especially excited about these aerological charts. He turned to the Saxony Ministry for Culture to obtain extra funding to continue production[32] and willingly gave up plans for including prac-

29. V. Bjerknes to H. Bjerknes, 16 January 1913, BFC, describes these events; his lecture, *Die Meteorologie als exakte Wissenschaft: Antrittsvorlesung gehalten am 8. Januar 1913 in der Aula der Universität Leipzig* (Braunschweig, 1913), 1–16, is translated as "Meteorology as an Exact Science," *MWR*, January 1914, pp. 11–14. On science and broader cultural trends in Wilhelmian Germany, see Russell McCormmach, "On Academic Scientists in Wilhelmian Germany," *Daedalus* 103, no. 3 (1974), 157–171; Lewis Pyenson, *Neohumanism and the Persistence of Pure Mathematics in Wilhelmian Germany*, Memoirs of the American Philosophical Society, 150 (Philadelphia, 1983).

30. Bjerknes to Woodward, 6 May 1913, copy, VBP; Bjerknes to Arrhenius, 8 May 1913, SAP.

31. The institute's publications were divided into two series: the first being the charts based on the international aerological observations, "Synoptische Darstellung atmosphärischer Zustände," and the second series being theoretical, methodological, and practical studies carried out at the institute, "Spezialarbeiten aus dem Geophysikalischen Institut."

32. Bjerknes to Arrhenius, 8 May 1913, SAP; Bjerknes to Sem Sæland, 16 November 1913, SSP, claims that he intends to push as far as possible to get funds by using the interest in establishing new aerological stations as a reason for expanding the series; in a letter to Wladimir Köppen, 14 May 1913, University Library, Graz, Bjerknes inquires about the costs of aerological institutes so he can justify the expensive maps in terms of saving through use of rational methods. He subsequently received 10,000 marks per year for publications (*Report of the Proceedings of the Seventh Meeting of The International Commission for the Investigation of the Upper Air: Held in Bergen 25th–29th July 1921*

tical aerological experiments at the institute in exchange for funding. This enthusiasm arose from the publication's ability to serve two functions in Bjerknes's plans. On the one hand, they demonstrated convincingly how his methods could be used practically for diagnosing atmospheric states; Bjerknes could use these charts to claim full legitimacy for his project as standard bearer for a dominating school. On the other hand, Bjerknes and his assistant could use these diagnoses to study the changes of three-dimensional atmospheric conditions from day to day to comprehend the laws of atmospheric change and to test prognostic methods, which after all would be the ultimate proof of his project's fruitfulness. Bjerknes had claimed in his inaugural lecture that it might take months to calculate a several-hour change in atmospheric state, but if the results agreed with the observed change, that would be a victory for science. But he and his group were equally interested in creating methods that could be put to practical use as soon as possible.[33]

To ensure close contact with practical aerology and the inclusion of practical techniques and methods at the institute, Bjerknes had written in advance to Hergesell requesting advice on an assistant.[34] In setting up the institute, Bjerknes turned frequently to Assmann for advice on practical aerology, a subject Bjerknes himself expected to become increasingly involved with because of its didactic value when teaching his methods.[35] He was made scientific advisor to the Leipzig Society for Aeronautics and was a guest of honor along with other dignitaries, including the king of Saxony and Graf Zeppelin, at the opening of the giant Zeppelin-airship hangar in Leipzig in June 1913.[36]

To develop prognostic methods based on atmospheric dynamics, Bjerknes and his assistants studied several subproblems, especially the definition and role of frictional forces. Bjerknes conceived the basic problem of dynamic prognosis as follows: if we have a chart of motion for three o'clock, is it possible to derive a chart for the conditions at six o'clock? He dismissed one possible method, comparing many such

---

[Bergen, 1921], p. 40); see reviews in *MZ* 30 (1913), 460–61; *Geographischer Monatsbericht,* September 1914, p. 144.

33. See, for example, in *Dynamic Meteorology and Hydrography,* pt. 1, "Remarks on the Rapid Work Essential for Daily Weather Service," p. 96, and in pt. 2, "Practical Applications of the Charts of Motion," p. 172.

34. Bjerknes to Hugo Hergesell, 22 June 1912, copy, VBP.

35. Bjerknes to Richard Assmann, 29 January, 25 February 1913, copies; Assmann to Bjerknes, 8 May 1913; all in VBP.

36. V. Bjerknes to H. Bjerknes, 15 June 1913, BFC; "Leipziger Verein für Luftschiffahrt," *DZL* 17 (1913), 99–100; "Eröffnung des Luftschiffhafens in Leipzig," ibid., p. 344.

charts and then deriving empirical rules for the changes observed, just as meteorologists use rules to forecast such changes in the pressure field as the movement of low- and high-pressure systems. Such "purely empirical methods can have only limited application. None of them are rational in a deeper sense of the word."[37] The distribution of pressure at a given time alone does not determine the pressure at a future time; similarly the wind field alone will not intrinsically determine a future field of motion. Only by recognizing the interaction between the pressure and motion fields as comprehended with the laws of hydrodynamics could a rational prognosis be conceived. Motion and pressure at three o'clock could, with the help of dynamic principles, yield the field of motion at six o'clock. From the new chart of motion, it should be possible to derive charts of vertical motion, which could then be correlated with cloudiness, precipitation, clear sky, and so forth. Of the various forces that needed to be taken into account, those related to friction were as yet the most problematic.

Back in Christiania, Hesselberg and Sverdrup had begun investigating the influence of friction on atmospheric motions; they continued these studies in earnest in Leipzig.[38] They tackled the problem of frictional forces from two perspectives. Given observations of the field of motion at a given time and at a slightly later time (approximately three hours), and knowing some of the forces producing this change (gradient of pressure, deflection caused by the earth's rotation), they attempted to find the missing force needed to produce the observed changes. They discovered that this complex force—called "friction"—could be treated as fairly constant average values. Once such average values were determined, the problem was then reversed: given a field of motion, to determine the change over the next three hours, considering all the forces involved, including friction, as knowns. Although in such calculations average values could not really succeed in replacing actual local and momentary values of friction, according to Bjerknes these attempts could nevertheless provide yet further insight into the nature of "atmospheric friction."[39] This and similar theoretical background work had to be resolved before *Dynamics* could be written.

Bjerknes's Leipzig school produced in its first years several valuable studies, some of which have become classics in the history of meteorology

37. Bjerknes, "Synoptical Representations of Atmospheric Motions," *QJRMS* 36 (1910), 283.

38. T. Hesselberg and H. U. Sverdrup, "Die Reibung in der Atmosphäre," *VGL*, vol. 1 (Leipzig, 1914), 241–319; T. Hesselberg, "Ueber eine Beziehung zwischen Druckgradient, Wind und Gradientenänderungen," Ibid., vol. 2 (1915), no. 7.

39. *CIWY, 1916*, p. 363.

and oceanography.[40] These included investigations of the influence of mountains on the motion of air close to the earth's surface;[41] thermal inversions in the "free" atmosphere;[42] a scale theory for atmospheric motions that would allow neglecting small-effect terms in hydrodynamic and thermodynamic equations;[43] applications of thermodynamics to atmospheric processes, by constructing tables for calculating the energy and entropy of moist air[44] and by studying the atmosphere and sea as heat engines;[45] internal waves and oscillations in the atmosphere;[46] and a dynamic and thermodynamic analysis of the trade winds in the North Atlantic based on aerological observations collected during an international experiment.[47] Further work included attempts to improve instrument design and measurement techniques and practical analyses of aerological data and aeronautical-meteorological problems for Assmann's and Hergesell's institutions.[48] Bjerknes lectured on statics and kinematics and then began on thermodynamics and dynamics in order to systematize materials and analyses for the remaining volumes of his *Dynamic Meteorology and Hydrography*.[49] All the investigations fed into this project and led to several revisions of the organization of *Dynamics*. By 1917, Bjerknes and his Carnegie assistants Sverdrup and Johan Holtsmark (who replaced Hesselberg in 1916) began work on dynamic prognoses.

40. I have decided not to consider these in any detail, in part because of space considerations and in part because they did not contribute directly toward the initial work of the Bergen school (except for Jacob Bjerknes's study considered below). Only when Bjerknes resumed theoretical work in 1920 did he return to problems begun in Leipzig.

41. T. Hesselberg and H. U. Sverdrup, "Ueber den Einfluss der Gebirge auf die Luftbewegung längs der Erdoberfläche und auf die Druckverteilung," *VGL*, vol. 1 (1914), pp. 101–16.

42. H. U. Sverdrup, "Ausgedehnte Inversionsschichten in der freien Atmosphäre," ibid., 75–100.

43. T. Hesselberg and A. A. Friedmann, "Die Grössenordnung der meteorologischen Elemente und ihrer raümlichen und zeitlichen Ableitungen," ibid., 147–73; Friedmann came from Russia to learn Bjerknes's methods so he could introduce them back home.

44. H. U. Sverdrup, "Der feucht-adiabatische Temperaturgradient," *MZ* 33 (1916) 265–72.

45. Bjerknes, "Ueber thermodynamische Maschinen, dei unter Mitwirkung der Schwerkraft arbeiten," *Abhandlungen SGW* 35, no. 1 (1916), 1–33.

46. Bjerknes, "Ueber Wellenbewegung in kompressiblen, schweren Flüssigkeiten," ibid., no. 2, 33–65; idem, "Theoretische-meteorologische Mitteilungen," *MZ* 32 (1915), 337–43.

47. H. U. Sverdrup, "Der nordatlantische Passat," *VGL*, vol. 2 (1917), 1–94.

48. H. Hergesell and Bjerknes, "Robert Wenger: Ein Nachruf," *BPfA* 10 (1922), v–viii; Bjerknes to Wiener, 4 August 1914, copy, VBP; Hesselberg to Bjerknes, 1 November 1915, JBP.

49. Bjerknes to Arrhenius, 3 June 1915, 25 August 1916, 25 March 1917, SAP; Bjerknes to Walfrid Ekman, 5 August 1915, VWE.

Crucial to the Bjerknes group's post-1917 work were the lines-of-convergence, which also reveal how issues important for aeronautics were incorporated in the institute's research.

Lines of Convergence

Bjerknes's Swedish assistant Sandström first noted lines of convergence around 1904, in the early kinematic investigation of currents and wind fields.[50] By drawing streamlines, which show lines of flow in a horizontal field of motion at a given time, Sandström found lines—subsequently called lines of convergence—where wind appeared to converge from two directions. Similarly, he noted lines where wind appeared to diverge into two horizontal directions—lines of divergence. These were only two of several geometric patterns discovered in the lines of wind flow. In accordance with Bjerknes's program of creating an exact physics of the atmosphere, such configurations were analyzed in these early studies as purely hydrodynamic phenomena requiring kinematic and dynamic explanation. Lines of convergence only began receiving great attention once Bjerknes and his assistants moved to Leipzig. There, these constructs figured in a number of studies on the influence of friction on atmospheric motions, and more significant, they became objects for study in their own right, being the problem given to the first doctoral student at the Leipzig Geophysical Institute, Herbert Petzold.[51]

Because Petzold's investigation has been regarded as one of the first contributions toward the post-1917 Bergen meteorology, it is instructive to know why lines of convergence became an object of inquiry at this time. In *Kinematics* (1911) there was a suggestion of a possible connection between lines of convergence and line squalls, the latter a weather phenomenon becoming increasingly important because of aviation.[52] A line squall was known to be a long (sometimes hundreds

50. J. W. Sandström, "Windströme im Gullmarfjord," *Svenska hydrografisk-biologiska kommissions skrifter* 2 (1905); *Dynamic Meteorology and Hydrography* 2:45–55.

51. Bjerknes and collaborators, *Hydrodynamique physique avec applications à la météorologie dynamique* (Paris, 1934), 3:847; Bjerknes to KVA's Nobelkomité för fysik, 26 January 1928, Nobel archives, Royal Swedish Academy of Sciences (KVA) (Bjerknes's nomination of T. Bergeron, J. Bjerknes, and H. Solberg for a Nobel physics prize, which was disqualified on statutory grounds: a prize cannot be divided equally among three persons).

52. *Dynamic Meteorology and Hydrography* 2:60. Many articles on line squalls began appearing around 1910 in both aeronautical and meteorological journals. For example, see the work of French meteorologists Paul Descombes, "L'atténuation des bourrasques pour la sécurité des aéronautes," *Revue aérienne*, 4th year, vol. 76, no. 73 (October–December 1911), 540–45, 633–34; and E. Durand-Gréville, "La loi des grains et la navigation aérienne," *L'aéronaute*, 44th year, n.s., no. 569 (7 January 1911), 4–12. In

of kilometers), rapidly moving, narrow band of thunderstorms or heavy showers, accompanied by powerful wind, which often arrived without much warning. Damaging winds and hail associated with strong line-squall thunderstorms had always been a problem in south and central European agricultural regions. Line squalls assumed even greater significance because of their devastating effects on balloons, airships, and airplanes.

As I have already noted, when the Leipzig Institute opened, a number of German states had established special observation networks for detecting line squalls and for sending warnings to aviators.[53] Commercial Zeppelin flights between German cities, including Leipzig, had already commenced, and military airship fleets had grown to major proportions.[54] For safe operation of airships, line squalls, had, at the very least, to be detected and tracked, preferably, to be predicted. In September 1913, several months after Bjerknes arrived in Leipzig, news of the destruction of the Imperial Navy's first military Zeppelin shocked Germany. The huge airship, for many, a symbol of German national supremacy, had been destroyed while on maneuvers by a powerful line squall over the North Sea.[55] Line squalls had ripped Zeppelins apart on the ground and wrecked smaller nonrigid airships; now a weather-related accident had resulted in the first loss of life in a Zeppelin crash. Here, then, was a problem for both aviator and meteorologist.

As scientific advisor to the Leipzig Society for Aeronautics and while keeping close contact with the Lindenberg aerological obser-

---

Britain, Napier Shaw and his assistants devoted considerable attention to line squalls; see especially R. G. K. Lempfert and Richard Corless, "Line Squalls and Associated Phenomena," *QJRMS* 6 (1910), 135–170. At the request of the Advisory Committee for Aeronautics, Shaw, a member of the committee, wrote a book on weather forecasting that included much of this work on line squalls (*Forecasting Weather* [London 1911], pp. 235–55) and prepared special reports for the committee on squalls and wind gusts; see also Lord Rayleigh, *Report of the Advisory Committee for Aeronautics for the Year 1911–1912* (London, 1912), p. 22; *Fifth Annual Report of the Meteorological Committee to the Lords Commissioners of His Majesty's Treasury: For the Year Ended 31st March 1910* (London, 1910), pp. 12–13; W. N. Shaw, "The Structure of the Atmosphere and the Texture of Air Currents in Relation to the Problems of Aviation (from Two Lectures at the Royal Institution 18 and 25 May 1911)," *Science Progress* 6 (1912), 345–47, 365–71. Franz Linke, *Aeronautische Meteorologie* (Munich and Berlin, 1911) 2:80–95, claims that line squalls are the greatest weather danger for aviators.

53. Wilhelm Peppler, "Die praktische Meteorologie im Dienste der Luftschiffahrt," *Wetter* (September–October 1909), 213–16, 234–39; Richard Assmann and Franz Linke, "Gewitter- und Böenwarnungsdienst," *DZL* 17, no. 9 (1913), 204–06.

54. Douglas H. Robinson, *Giants in the Sky: A History of the Rigid Airship* (Seattle, 1973), pp. 40–84.

55. Douglas H. Robinson, *The Zeppelin in Combat* (London, 1962), pp. 24–26; L. Deighton and A. Schwartzman, *Airshipwreck* (New York, 1978), p. 20.

vatory, Bjerknes kept abreast of the numerous meteorological problems confronting the ever-expanding aeronautical activities in Germany.[56] Soon after arriving in Leipzig, he discussed with Assmann general problems in aerology, operations at Lindenberg, and possible cooperation between their respective institutions.[57] Bjerknes had after all promised in 1911 to help Assmann with his line-squall warning system for aviators. Not surprisingly, then, the first doctoral project at the institute was on the relationship between line squalls and lines of convergence—an attempt to comprehend kinematically this important weather phenomenon.

Petzold investigated the problem for one year, before he was recruited into the army and subsequently killed at Verdun. His preliminary studies strengthened the belief in a relationship between lines of convergence and line squalls. He found that near the ground, squalls were accompanied by three parallel horizontal lines, one after one another: a line of convergence in the wind field, a trough line in the pressure field, and a line across which occurs the fastest temperature fall.[58] During the war, when aerial operations increased dramatically, line squalls became an ever more serious problem; many articles and reports relate attempts to locate, comprehend, and predict them.[59] In 1917, Bjerknes's eighteen-year-old son, Jacob, resumed this investigation. While studying natural science and mathematics at Leipzig, Jacob had joined his father's project, which at the time seriously lacked assistants.

By this time, Bjerknes, like all meteorologists, understood that their discipline was undergoing major changes. The war was not only disrupting his institute; it was altering the weather's social significance to a degree not previously imaginable. Based on his earlier experiences and on unexpected contingencies and exigencies, Bjerknes acted on these changes.

56. See meeting reports of the Leipziger Verein für Luftschiffahrt in *DZL*.

57. Assmann to Bjerknes, 31 January, 8 May 1913; Bjerknes to Assmann, 25 February 1913; all in VBP. For the earlier correspondence, see Assmann to Bjerknes, 16 May 1911, which refers to the contents of Bjerknes's letter of 6 May; Bjerknes wrote a draft for his reply, 23 May, on Assmann's letter.

58. Related in Jacob Bjerknes, "Ueber die Fortbewegung der Konvergenz- und Divergenzlinien," *MZ* 34 (1917), 349; and in V. Bjerknes to Nobelkomité, 26 January 1928, Nobel archives, KVA.

59. G. P. Neumann, "Der Heereswetterdienst," *Die deutschen Luftstreitkräfte im Weltkriege* (Berlin, 1920), pp. 292–93; P. Schereschewsky, "La météorologie militaire pendant la guerre," *L'aéronautique* 1 (1920), 532–33; almost all meteorological and general aviation journals of the period contain discussions of the line squall problem for aerial operations.

PART III

WARTIME EXIGENCIES: PRACTICAL FORECASTING IN BERGEN AND THE ORIGINS OF A NEW CYCLONE MODEL (1917–1918)

[5]

Bjerknes Turns to
Weather Forecasting

WORLD WAR I transformed meteorology. Directly and indirectly, the war provoked change in the goals, methods, and organization of weather forecasting; after 1914, weather began to be perceived and used differently. Neither Bjerknes nor his institute remained unaffected. Bjerknes reluctantly returned to Norway in response to a call to the Bergen Museum in 1917. In attempting to overcome the uncertainties of working once again on the periphery and of having too few resources for completing his project, Bjerknes altered the scope of his endeavors to include practical weather forecasting, which in turn brought unexpected results.

Bjerknes's call to a professorship at the Bergen Museum stemmed from a plan to establish a faculty for mathematics and natural sciences, a step toward developing a new university. Bergen had been a major trading and cultural center in Norway since the Middle Ages; local patriotism among her leading citizens was also an old tradition. Having been central to Bergen's scientific and cultural life for nearly a century, the museum was now poised to expand into the long-awaited university for West Norway. Although serious discussion and fund raising for erecting a university had begun in the 1890s, little progress had been made before the war. In 1915 the museum's board of directors renewed the drive to create the college and ultimately a university. Funds from private sources came easily during the early war years; huge profits from shipping enabled many wealthy Bergen patriots and patrons of the museum to pledge to the endeavor.[1]

1. *Bergens museum 1925: Historisk fremstilling* (Bergen, 1925), pp. 66–71. In 1916, Bergen shipping earned, net, 104 million crowns, as compared with 47.7 million in

Helland-Hansen, the museum's first professor (1911), proposed in 1916 that an institute for geophysical science be established to serve as a cornerstone for the proposed college. For more than a decade Helland-Hansen and members of the museum's biologic station had been offering a series of courses in oceanography and marine biology for Norwegian and foreign students; the investigations of the diverse Bergen oceanographic researchers had achieved broad international recognition.[2] After discussing the issue with Bjerknes and other geophysicists, Helland-Hansen recognized that a geophysics institute could bring together researchers from the various fields in which Norwegians could claim achievements: meteorology, oceanography, and physics of the earth (terrestrial magnetism and northern lights). Bringing these specialties together in one institute could stimulate yet greater scientific and practical results.[3] As a member of the museum's board of directors, he readily gained support. Within a few months enough private money had been collected to guarantee the founding of a geophysics institute, although at first only oceanography and meteorology would be represented.[4]

When Helland-Hansen informed Bjerknes of the museum's decision to create the institute, he also tried to interest Bjerknes in coming to Bergen. Bjerknes politely wished the new endeavor success and repeated his earlier statements that in principle he had nothing against coming to Bergen, but only after he had established a vital school of dynamic meteorology in Germany.[5] Several weeks later, Bjerknes was surprised by a telegram claiming that a personal professorship at the museum had been established for him. As he said, "A number of persons intervened in my life."[6]

Bjerknes and his family had been managing relatively well during the war by receiving food shipments from Norway and by spending summer holidays there. As food shortages began to develop in Nor-

---

1915 and with an average of 10 million during the years 1911–14. Toward the end of 1916, losses increased because of stepped up German U-boat actions against neutral ships; nevertheless, profits were astronomical, and investment speculation was conducted at a feverish pace. Per Vogt, *Jerntid og jobbetid: En skildring av Norge under verdenskrigen* (Oslo, 1938), p. 112; J. H. Lyshoel, "War Time Business in Norway," *American Scandinavian Review* 4 (1916), 220; Berge Furre, *Norsk historie, 1905–1940* (Oslo, 1972), pp. 91–96.

2. Bjørn Helland-Hansen and August Brinkmann, "Biologisk station," in *Bergens museum 1925*, pp. 417–54.

3. Ibid., pp. 45, 462–63; see documents in KUD/D, folder: "Bergens museum: Geofysisk institutt, Budsjett 1918/1919".

4. Bergens museum styre (J. Lothe, C. F. Kolderup) to Bjerknes, 17 March 1917, VBP.

5. Bjerknes to Bjørn Helland-Hansen, 18 February 1917, BHH.

6. Bjerknes to Sem Sæland, 21 November 1923, SSP.

way as well and as diplomatic relations between Germany and Norway worsened, food shipments were first restricted and finally, after January 1917, no longer permitted. One last shipment to the Bjerknes family was granted, however, in February, through government intervention.[7] Fearing for her son's health, Bjerknes's mother, Aletta, called upon Nansen and asked for assistance in getting Vilhelm back to Norway as soon as possible.[8] Aletta Bjerknes feared her son's health might collapse well before the summer vacation, when he could come to Norway.

Nansen wrote immediately to Helland-Hansen. Fearing the worst, and recognizing that they would not easily forgive themselves if they did not act quickly to help, Nansen wondered whether Helland-Hansen could not arrange immediately a position for Bjerknes at the new geophysical institute. Bjerknes could help in organizing the institute and other facilities while benefiting from the better living conditions. Though it might be possible to reestablish Bjerknes's old personal professorship at the university in Christiania, Nansen considered the Bergen option more expedient. Helland-Hansen acted quickly. He called the museum's president, Johan Lothe, and asked if they could gather together fifty thousand crowns to cover Bjerknes's salary and expenses for an initial three years. Within a few hours, everything was arranged. Shortly thereafter the board of directors voted to call Bjerknes to a professorship entailing full freedom to do as he chose, to be taken up as soon as possible.

When the telegram arrived in Leipzig, Bjerknes was at a resort in the German Alps gathering strength, preparing lectures, and skiing.[9] Although somewhat perplexed by the tones of crisis and by the unexpected suddenness of the call, Bjerknes nevertheless expressed cautious interest.[10] True, the food and fuel shortages had become acute; still, his loyalty to the institute and to his project dominated his thinking. He was also well aware of the war's continued impact on his institute and its work.[11] Most of his students and assistants had been called into military service, and most of them had been killed.[12] Through contacts at the Saxony Cultural Ministry, Bjerknes had

7. See letters from Kristine Bonnevie, January–April 1917, on efforts to send food to the Bjerknes family in Leipzig, BFC.

8. Fridtjof Nansen to Helland-Hansen, 13 March 1917, in, *Fridtjof Nansen Brev*, ed. S. Kjæheim, vol. 3 (Trondhjem, 1963), p. 272; Liv Nansen Høyer, *Nansen og verden* (Oslo, 1955), pp. 132–33.

9. V. Bjerknes to H. Bjerknes, 17, 23 March 1917, BFC.

10. Ibid., 23 March 1917; V. Bjerknes to Aletta Bjerknes, 24 March 1917, BFC; V. Bjerknes to Bonnevie, 24 March 1917, BHH.

11. V. Bjerknes to Walfrid Ekman, 2 September 1917, VWE.

12. Bjerknes to Svante Arrhenius, 10 May, 3 November 1916, SAP.

managed to retain his primary German assistant, Robert Wenger, but in a major expansion and reorganization of the military weather service at the end of 1916, Wenger's expertise could no longer be sacrificed, and he too was called in.[13]

Still more threatening for the continued progress of the project, Bjerknes's most experienced Norwegian Carnegie assistants were departing. Hesselberg became in 1916 head of the Norwegian Meteorological Institute, and Sverdrup was planning to become the scientific leader of Roald Amundsen's polar expedition on the *Maud*. Although he was able to recruit younger Norwegians (Johan Holtsmark, Halvor Solberg, and eventually his son Jacob), they required time to become fully acquainted with the intricacies of the project. By 1917, Bjerknes had to acknowledge that the loss of so many experienced assistants and most of his doctoral students seriously restricted the institute's research endeavors.[14] Earlier he had claimed that although it might be convenient to leave in a time of crisis, he would not dare jeopardize everything he had worked for. A difficult decision confronted Bjerknes.

For his part, Helland-Hansen began to do everything possible to entice Bjerknes to Bergen. He informed Bjerknes that all planning for the institute would wait until he arrived; he even suggested that the institute could regularly publish synoptic charts for various levels in the atmosphere—a series indentical to that which was a cornerstone of the Leipzig institute.[15] Bjerknes still could not imagine simply abandoning his institute. He sought to have Wenger appointed professor and in this manner to keep the institute on the path he had set it on. He tried to persuade the dean and his Leipzig colleagues that the only way the institute's research program and methods could achieve a final, polished form would be for him to continue with them in Bergen rather than continuing in wartime Leipzig "under the worst possible conditions."[16] Not knowing whether the relatively young Wenger could be appointed professor and whether he would even return from the front fit for such duties, Bjerknes wanted a one-year leave of absence, in the hope that the situations in Leipzig and in Bergen would become clearer. Regardless, Bjerknes did not want to leave until the semester ended; for he did not share his mother's

13. Bjerknes to Helland-Hansen, 18 February 1917, BHH; H. Hergesell and V. Bjerknes, "Robert Wenger: Ein Nachruf," *BPfA* 10 (1922), vi.

14. Bjerknes, "Hvordan Bergenskolen ble til," in *Vervarslinga på Vestlandet 25 år: Festskrift utgitt i anledning av 25-års jubileet 1. juli 1943* (Bergen, 1944), 11–12; Bjerknes had himself engineered the Hesselberg appointment; Sverdrup, his other main Carnegie assistant, gave notice.

15. Helland-Hansen to Bjerknes, 26 April 1917, copy, BHH.

16. Bjerknes to Helland-Hansen, 16 April 1917, BHH.

extreme fears about his health. Finally, he was refused a leave of absence; Wenger, however, was to be called in from the front and to be appointed professor.[17] Bjerknes reluctantly accepted the Bergen call and resigned his position in Leipzig.[18]

Upon returning to Norway during the summer of 1917, Bjerknes nevertheless expressed great relief and happiness. When sitting at the Fløien restaurant, on one of the seven mountains ringing Bergen, he exclaimed: "When I look out over all this, I have to think: This *is* Norway."[19] If he was to suffer the indirect effects of the war, he would prefer to suffer among his own people.[20] But he was also wary; twice before he had been forced to leave Norway by the poor conditions for research and lecturing. Even before arriving in Bergen, Bjerknes learned that his situation would not be as promising as Helland-Hansen had claimed: housing and office space were extremely scarce and expensive, especially after the massive Bergen fire in 1916, and soaring inflation made it unreasonable to expect new construction for the institute to begin soon. Bjerknes was offered temporary work space in the museum's oceanographic laboratory.[21] Bjerknes soon began considering how to improve the chances for success of his research and professional goals in Bergen. To avoid the pitfalls of geographic isolation and scarce resources and attain his goal of transforming meteorology as a science and a profession, Bjerknes required a vital national discipline.

While resting in the mountains, near Gjeilo, Bjerknes gathered together leading Norwegian geophysicists and his former Swedish assistant, Sandström, to discuss the creation of a geophysics association.[22] In conjunction with the association, they established a commission to advise and lobby the Ministry of Church and Education on matters relating to research, institutions, and training in the geophysical sciences. Bjerknes recognized that these plans for organizing Norwegian geophysicists into an interest group were "quite ambitious"; he also hoped later to create a physics society, which similarly could

17. Ibid., 13 June 1917, BHH.
18. H. Bjerknes to A. Bjerknes, 25 April 1917, BFC; H. Bjerknes to Helland-Hansen, 31 May 1917; V. Bjerknes to Helland-Hansen, 13 June 1917; both in BHH.
19. V. Bjerknes to H. Bjerknes, 1 September 1917, BFC.
20. V. Bjerknes to Arrhenius, 23 August 1917, SAP.
21. Helland-Hansen to Bjerknes, 27 June 1917, copy, BHH.
22. "Aar 1917 18de–20de august holdt undertegnede et møte paa Gjeilo og besluttet følgende . . . ," copy, JBP, cosigned by V. Bjerknes, O. Devik, B. Helland-Hansen, T. Hesselberg, Ole Krognes, H. U. Sverdrup, Sem Sæland; letters of invitation to join were sent also to Nansen, Carl Størmer, B. J. Birkeland, and Lars Vegard; see also T. Hesselberg, "Organisationen av det meteorologiske arbeide i Norge," NMIÅ *1919–1920*, pp. 4–6; Bjerknes to Nansen, 16 September 1918, FNP.

"appear in public like the geophysics association in order to care for physics' needs."[23]

To promote Norwegian geophysics both at home and abroad, the group established a fund for geophysical research. By naming the fund after the recently deceased Kristian Birkeland (who, in addition to his work on northern lights, had also contributed to the pioneering artificial-nitrate process used in establishing Norsk Hydro), they appealed to Norway's industrial-commercial leaders for contributions to advance the highly practical geophysical sciences.[24] Interest from this fund would support research and a new journal, *Geofysiske publikasjoner* for publicizing Norwegian geophysical contributions. Rather than depending upon the goodwill of foreign journal editors, they would publish themselves. By accepting contributions in English, French, and German and by publishing in a neutral country, they could direct their results to both sides of the war-torn scientific world. In Leipzig, Bjerknes had recognized the need for a publication series to circumvent skeptical older authorities; obviously the same strategy would apply in Bergen.

Of greatest immediate importance for Bjerknes, of course, was how to resume the Leipzig research program. He was determined to complete volumes 3 and 4 (on dynamics and thermodynamics) of his *Dynamic Meteorology and Hydrography* and in the process achieve rational precalculations of atmospheric phenomena.[25] In spite of the worsening conditions in Germany, Bjerknes's last year there had been "very fruitful": "We have made for the first time progress on meteorological predictions based on dynamic principles."[26] Although it was too early to assess the practical significance of these developments, Bjerknes considered them theoretically promising. He and his assistants had used hydrodynamic equations on an analysis of a wind and pressure reading to try to calculate the wind field three hours later. They failed, according to Bjerknes, because their wind data did not represent true atmospheric conditions. Various local influences combined with the extreme complexity of friction near the ground distorted the wind sufficiently from theoretical ideal conditions to ruin the calculation.[27]

If they could obtain aerological observations undisturbed by ter-

23. Bjerknes to Arrhenius, 23 August 1917, SAP; Bjerknes to Sæland, 25 September 1919, SSP.

24. Helland-Hansen to Sam Eyde, 28 August 1917, BHH. In all the relevant documents, the authors emphasized the dovetailing of scientific and practical interests for all the geophysical sciences.

25. Bjerknes to Arrhenius, 23 August 1917, SAP; Bjerknes to R. S. Woodward, 16 August 1917, draft in Carnegie file, VBP.

26. Bjerknes to Arrhenius, 23 August 1917, SAP.

27. *CIWY, 1916–1917*, p. 322; Bjerknes's copy of the report is dated 6 August 1917.

rain, "it should probably be possible, not only to precalculate wind for short intervals of time with satisfactory accuracy, but also to produce general weather forecasts of practical value."[28] Bjerknes had been hoping that the U.S. Weather Bureau, with its large budget and its geographically broad observation network, could provide him just such data. But while still in the mountains during the summer, he came across an opportunity to obtain in Norway a regular supply of aerological data and—equally enticing—the trainees to put into practice the rational methods of analysis.

Bjerknes discovered that the person living in the neighboring cabin was minister for defense, Theodore Holtfodt. Together they discussed German aeronautical and military capabilities.[29] Bjerknes reported that modern warfare and military defense now included specialized weather services; perhaps Norway's growing defense also required such meteorological facilities? At Holtfodt's request, Bjerknes began a report on the German field weather service and the feasibility of a comparable service for Norway.

Meteorology, World War I, and Norway

In Germany during the war Bjerknes had followed the extraordinary growth of weather's role in warfare and of the military weather forecasting services. He had little choice, for although he himself remained neutral, his project and institute did not have the luxury of remaining outside the conflict.[30] His and his assistants' work was exploited by the German military.[31] The military weather service exhausted the available supply of the German edition of *Dynamic Meteorology and Hydrography* (even the American edition was in short supply) and often requested publications from the institute.[32] Military meteorologists came to Bjerknes for information and training; analytic methods devised by Bjerknes's assistants were adopted in the field weather service.[33] Through direct discussion with German col-

28. Ibid., p. 323.
29. My discussion is based on the following materials: V. Bjerknes to H. Bjerknes, 22 August, 4 September 1917, BFC: V. Bjerknes to Theodore Holtfodt, 19 September 1917, copy, SSP and VpVA; V. Bjerknes to Holtfodt, 18 February 1918, copy, VpVA.
30. Bjerknes to Arrhenius, 2 August 1916. SAP.
31. H. Bjerknes to M. & S. Arrhenius, 27 December 1916, SAP; V. Bjerknes to F. H. Wedel-Jarlsberg, 29 March 1920, copy, VBP.
32. Friedrich Vieweg & Sohn to Bjerknes, 2 December 1919; Bjerknes to Friedrich Vieweg & Sohn, 3 December 1919, copy; Woodward to Bjerknes, 14 June 1918; Bjerknes to Woodward, 24 August 1918, copy; all in VBP.
33. Noted in Bjerknes's "confidential" letter to Minister for Defense Holtfodt, 19 September 1917, copy, SSP and VpVA; H. Bjerknes to M. & S. Arrhenius, 27 December 1916, SAP, describes the German military's interest in Bjerknes's institute and project.

leagues, and through letters from his students, Bjerknes followed the challenges faced by meteorologists at the front. His knowledge of meteorology during the war thus helped shape Bjerknes's subsequent proposals and actions.

During World War I, mechanization of warfare and rationalization of military and economic activities created a demand for increased knowledge of the atmosphere and for improved means of weather prediction.[34] Although many sciences were called upon to work for the war, the enormous growth in meteorology's social significance was especially noticeable: "There is no more interesting illustration of the application of new scientific methods to warfare than is furnished by the developments in meteorology during the great war. Prior to 1914 a meteorological section was not considered a necessary part of the military service."[35] Toward the end of the war each combatant's army, navy, and airforce possessed specialized field weather services that provided forecasts and data for planning strategies. Preexisting meteorological institutions greatly expanded their operations both to coordinate the various specialized field weather services and to assist war-related activities on the home front.

Recognizing that the weather would be a factor in waging war, German military planners had organized weather services for aerial operations before war broke out.[36] Few elsewhere had yet recognized the weather as a critical military concern: "As far as the [British] army went . . . there was practically no provision for any meteorological information in 1914, the official attitude being that it was unnecessary; British soldiers, unlike the Chinese, did not go into action carrying umbrellas."[37] The French, too, had considered meteorology gen-

34. Of the seemingly endless body of articles, reports, and monographs on the development of each nation's military weather services, a sample of those I consulted includes: Werner v. Langsdorff, "Luftbildwesen, Funkentelegraphie und Wetterkunde in ihrer Bedeutung für die Luftwaffe," in *Unsere Luftstreitkräfte, 1914–1918*, ed. Walter von Eberhardt (Berlin, 1930), pp. 86–90; G. P. Neumann, "Der Heereswetterdienst," *Die deutschen Luftstreitkräfte im Weltkriege* (Berlin, 1920), pp. 286–97; C. Leroy Meisinger, trans., "Notes on the Meteorological Service in the German Army from Translations of German Documents," *MWR* 47 (1919), 871–74; H. C. Lyons, "Meteorology during and after the War," ibid. 47 (1919), 81–83; Alexander McAdie, "Meteorology and the National Welfare," *Scientific Monthly* 6 (1918), 176–86; Ernst Gold, "The Meteorological Office and the First World War," *Meteorological Magazine* 84 (1955), 173–78. William Napier Shaw, "Meteorology: The Society and Its Fellows," *QJRMS* 45 (1919), 95–111, discusses British meteorology's role in the war effort and the impact this had on meteorology as a science and a profession.

35. Robert A. Millikan, "Some Scientific Aspects of the Meteorological Work of the United States Army," *Proceedings of the American Philosophical Society* 58 (1919), 133.

36. Neumann, "Heereswetterdienst," pp. 286–87; "Germany's Aeronautical Weather Bureau," *Scientific American* 108, no. 12 (1913), 262; Richard Assmann and Franz Neumann, "Warnungsdienst für Luftfahrer," *DZL* 14, no. 26 (1910), 3.

37. C. J. P. Cave, "Some Notes on Meteorology in War Time," *QJRMS* 48 (1922), 7.

erally irrelevant to their combat operations, but in 1915 when poison gas attacks threatened to become decisive in the war, both French and British weather services were created behind the fronts.[38] Similarly, the bombing of towns on the English east coast in 1915 indicated to meteorologists that the Zeppelin airships were carried north and east of their obvious target—London—by upper winds of which the Germans had no knowledge. Shaw made the lesson clear to the military authorities: meteorology could not be ignored in the war.[39] Soon, increased allied use of airplanes and observation balloons also necessitated support from weather services in close proximity to the combat zones.

By 1917 and 1918, advances in airplane and airship construction led to an evolution in aerial warfare. Routine bombing and night flying required specialized aviation weather services, "to make aviation a more effective means of waging war, and to save the lives of some of our flyers."[40] Development of long-range artillery bombardments and use of barrage gunnery to cover advancing troops again required specialized weather services: "Meteorological data is absolutely essential to the successful operation of the artillery in modern warfare."[41] Meteorology's utility for almost all aspects of warfare had become obvious to meteorologists and military officials. Unlike chemistry, which could be exploited directly for the production of weapons (like poison gas), meteorology's function lay more in its promise to aid the rationalization of a massive war effort.

At the front, previously unexplored weather phenomena and rarely experienced forecasting problems confronted meteorologists. For the field weather services to support military operations, forecasts of location, time, and varieties of weather phenomena had to achieve

38. P. Schereschewsky, "La Météorologie militaire pendant la guerre: Historique sommaire," *L'aéronautique* 1 (1920), 532. For gas warfare's role in the setting up of British field weather services, see Cave, "Some Notes on Meteorology," p. 7; Gold, "Meteorological Office," pp. 173–75. On the relations between British science and military during the war, see R. M. Macleod and E. K. Andrews, "Scientific Advice in the War at Sea, 1915–1917: The Board of Invention and Research," *Journal of Contemporary History* 6 (1971), 1–40; M. Pattison, "Scientists, Inventors and the Military in Britain, 1915–1919: The Munitions Inventions Department," *Social Studies of Science* 13 (1983), 521–68.

39. Gold, "Meteorological Office," p. 173.

40. Robert de C. Ward, "Meteorology and War-Flying: Some Practical Suggestions," *Annals of the Association of American Geographers*, 8 (1918), p. 5.

41. U.S. Dept. of Agriculture and War Dept., Bureau of Aircraft Production, "Report of Activities of Meteorological Section, Signal Corps.: During the Year October 1917 to October 1918," Science and Research Dept. Reports, October 1918, National Academy of Sciences Archives, Washington, D.C. (I thank Richard Hallion for a copy of this report); see also, for example, Millikan, "Some Scientific Aspects," pp. 133, 143–45; E. M. Wedderburn, "Meteorology and Gunnery," *Journal of the Scottish Meteorological Society* 18 (1919), 86–92.

unprecedented precision and detail. "It was found that the demands
for forecasts in connexion with warlike operations were even more
insistent than those made in times of peace, and the forecasts required, particularly those for aviation and gas warfare, were of a
highly detailed character. Forecasts for such operations must be very
definite."[42] Military concerns directed attention to a three-dimensional atmosphere. Weather conditions well above the ground (for aviation and gunnery), including the degree and type of cloud cover had
to be stressed, as did small-scale atmospheric phenomena, such as
turbulence and eddy motions in the layers of air in nearest proximity
to the ground (for gas attacks and for aviation). The upper air was no
longer of interest simply to aerologists; virtually all meteorologists
had to consider phenomena aloft.[43] Common to all the efforts to
satisfying the new forecasting criteria was the establishment of field
weather services based on unusually dense networks of observation
stations and frequent, rapid exchanges of data. Forecasters devised a
wide range of ad hoc predictive methods. Meteorologists were only
too conscious of the tension between the state of their science and the
tasks it was called upon to perform.[44]

Having kept abreast of these developments, Bjerknes recognized
that conditions favored arguing for an expansion of Norway's weather services. Lack of an airforce had made widespread aerological
observations hitherto unnecessary and certainly impractical, given the
budgetary limitations of the Norwegian Meteorological Institute. But
the situation had changed, as Norway's neutrality was strained during
the war. Allied and Central powers vied for access to Norway's pre-
cious resources: fish, copper-laden pyrites, and her merchant fleet.[45]
Attempts by Britain and Germany to restrict Norway's trade in their
own favor proved equally problematic. German U-boats sank Nor-

42. Air Ministry, *The Weather Map*, 4th ed. (London, 1926), pp. 6–7.

43. Shaw, "Meteorology," p. 100; Gold, "Meteorological Office," p. 175; F. Linke, *Die meteorologische Ausbildung des Fliegers*, 2d ed. (Munich and Berlin, 1917), pp. 28–58; the necessity of studying the meteorology of the upper air is covered in virtually all the relevant literature.

44. Although the ad hoc forecasting methods were developed in response to wartime military meteorological needs, they generally entailed formalistic empirical rules. In this respect they perpetuated the prewar tradition whereby forecasting methods and physical understanding of atmospheric phenomena rarely interacted. Shaw ("Mete-orology," p. 102) remarks that wartime experiences had revealed the limits of such empirical methods; meteorology's increased social importance requires that "we must depend upon our knowledge of the dynamics and physics of the atmosphere"; this belief was also expressed by the French astrophysicist/meteorologist, Henri Deslandres, to H. H. Hildebrandsson, 26 January 1918, HHH. But how to find methods to use this knowledge remained a problem.

45. Olav Riste, *The Neutral Ally: Norway's Relations with Belligerent Powers in the First World War* (Oslo, 1965).

wegian merchant vessels; air and sea strikes against Norway were real possibilities.

The Norwegian government reacted by increasing the defense budget several fold to purchase military hardware and to mobilize a reserve army.[46] This military build-up included the formation of air forces in the army and navy. The Ministry of Defense had in 1912 first received appropriations for airplanes and pilot training.[47] During the early years of the war significant growth in the air force began, most notably the building of airplane factories at the chief army and navy air force bases (Kjeller in 1916 and Horton in 1915, respectively). To defend her neutral waters from U-boat warfare, the navy built or purchased (from Britain) several larger planes for maritime reconnoitering and bombing. In the spring of 1917 major organizational expansion occurred, "in that aviation's significance and necessity became clearly evident, on account of the then-prevailing military situation."[48] With developing air defenses, a field weather service, including pilot balloon observations of upper wind patterns, was justifiable.

Less than a month after speaking with Holtfodt, Bjerknes sent him a confidential report, which circulated under a secret classification within the Ministry of Defense.[49] In the report he described the expansion of the prewar German network of twenty aerological observation stations into an intricate, dense network of stations behind all the fronts. Surface and upper-air observations, made at least three times daily, were being exchanged rapidly between stations by wireless telegraphy and field telephone systems. With this data, a central station in each military district could prepare its own specialized local forecast. Established to serve Zeppelin and airplane flights, the field

---

46. Statistisk Sentralbyrå, *Norges offisielle statistikk XII 245: Historisk statistikk 1968* (Oslo, 1969), p. 450 (hereafter cited as *NOS XII 245*). Table 234, "Revenue and Expenditure of Treasury, 1913–14 to 1945–46," shows the following increase for expenditure on military and civil defense: 1913–14, 27.5 million crowns; 1916–17, 93.6 million crowns; 1917–18, 134.7 million crowns.

47. Frederik Meyer, *Hærens og marinens flyvåpen, 1912–1945* (Oslo, 1973), pp. 13–21; Nordic Pool for Aviation Insurance, *Luftfartsbok omhandlende Danmark, Finland, Norge og Sverige* (n.p., 1922), p. 151.

48. Nordic Pool for Aviation Insurance, *Luftfartsbok*, pp. 152–53; Meyer, *Hærens og marinens flyvåpen*, pp. 35, 43, 50; Torleif Lindtveit and Gunnar Thoresen, *På vingene over Norge* (Oslo, 1980), pp. 66–76, 83–90. For the great financial and psychological impact on Norway of stepped up U-boat warfare, see "Skibsfarten og ubåtkrigen," in Wilhelm Keilhau, *Norge og verdenskrigen*, Carnegie Endowment for International Peace: Economic and Social History of the World War, Scandinavian Series (Oslo, 1972); and Riste *Neutral Ally.*

49. RA contains Bjerknes's report and documents related to the Ministry of Defense's actions on the matter: Forsvarsdepartementet, journal no. 729, 2195/18, and folder: "Feltveirtjeneste."

weather services proved valuable for virtually all military planning. Bjerknes noted that in addition to assisting in gas attacks, aerial operations, and long-range artillery bombardments, these meteorological services were used to lift morale and save energy during troop movements and ground attacks. He conceded that weather would only rarely be decisive in war, but the value of weather services to military operations could be considerable.[50] A comparable service could be erected in Norway, but on a smaller scale.

Bjerknes did not plan to work actively in the proposed service, but he expected to benefit from it. He called for training some forty "field meteorologists" to be stationed at military posts around the nation and at locations selected because of their meteorological importance. These meteorologists were to take observations, including aerological ones, and send them by telegraph (eventually by wireless) to three central stations where higher ranking military meteorologists could analyze them and send forecasts back to the individual military posts. To lead this service Bjerknes recommended Sverdrup, who was then a reserve first lieutenant. Although Sverdrup planned to leave on Amundsen's forthcoming polar expedition, Bjerknes believed that his former assistant would be available through all of 1918 to set up the service and train the would-be meteorologists; in the process, he could introduce their rational methods. Bjerknes expected that students with scientific education who entered military service could absorb this specialized training in a several-week course. Bjerknes could soon gain access to Norwegian aerological data for experimenting with rational precalculations and a corps of trained practitioners.

Bjerknes also suggested to Holtfodt that direct interaction between a field weather service and an institute for dynamic meteorology could lead to great progress in developing a rational science of weather. His Leipzig institute had in fact already begun to focus on a "systematic scientific analysis" of the German field weather service's problems and observations. His new institute in Bergen would "gladly join in the processing of the [Norwegian] observational data," and he personally would be available to offer advice and information for a Norwegian field weather service.[51]

Bjerknes's proposal was more than an attempt to exploit a wartime situation. He was considering both the short-term requirements for

50. V. Bjerknes to Holtfodt, 19 September 1917, copy, SSP and VpVA; the perceived importance then of a military field weather service for Norway may be inferred from Sem Sæland's response to Bjerknes on this matter. Sæland, a physicist and member of the Storting's military committee, thought the idea so reasonable that funding would come very easily and even the socialists would probably go along (Sæland to Bjerknes, 8 September 1917, VBP).

51. Bjerknes to Holtfodt, 19 September 1917, copy, SSP and VpVA.

resuming his project and the long-term prospects for founding a new Norwegian meteorological school through which to further his methods and research. Moreover, as always, he understood that any national endeavor had to be conceived with international developments in mind. In Leipzig, Bjerknes was in the center of Continental activities and director of the most significant academic institute for aerology or, for that matter, meteorology. Obviously, he was still thinking about the future of international meteorology. His experiences in Germany had shown him that a new era for meteorology was beginning and that his original plans for developing a physics of the atmosphere perhaps needed rethinking.

Opportunities in Weather Forecasting: Agriculture and Aviation

During the fall 1917, Bjerknes acknowledged that the time had come for dynamic meteorology to develop into an autonomous science and in so doing to reclaim weather forecasting from the prevailing statistical-climatological approach. Forecasting's social significance and urgency were certain to grow. The evolution of military technologies and strategies suggested that weather services for the military would become permanent, war or no war.[52] Widespread commercial flying was now within reach and would depend on greatly expanded weather services. Wartime aviation showed forecasting required greater detail and precision, and not simply in aerological observations; here was an opening for Bjerknes to advance his claims for a meteorology based on physics.

In a growing international discipline dominated by men of conservative, empirical temperament, Bjerknes could not afford to be passive. Even more than in the pre–World War I period, he needed to show that his approach could best solve the problems confronting meteorologists, especially those problems whose solution would be vital for advancing their profession. He noted, "We are living in an age in which new demands on meteorology are going to be made on a grand scale, . . . above all through the coming air traffic"[53]; moreover, "the world is entering a period of food shortage, and in this time [dynamic] meteorology has a responsibility to support, as well as it utmost can, the production of food stuffs by developing and improving weather forecasting."[54] Although he had mentioned in the pro-

52. Ibid.

53. Draft of Bjerknes's letter of recommendation for Sandström to head a proposed meteorological institute in Gothenburg, n.d. [fall 1917], Sandström file, VBP.

54. Letter of recommendation for Sandström, n.d. [fall 1917], copy in Sandström file, VBP; see also Sandström to Bjerknes, 14 December 1917, VBP.

posal for a field weather service that the predictions could also be used to assist farmers during this time of growing food shortages, the threat of famine prompted Bjerknes in 1918 to work directly toward just this end.

In Aid of Agricultural Production

Worsening food shortages during the war compelled the Norwegian government to intervene in virtually all aspects of agricultural production. Shortages in grain and feed concentrate (the latter for animal consumption) began developing in 1916.[55] In that year, only 350,000 tons of grain were produced, which, compared with the 700,000–800,000 tons consumed during each of the first two war years, caused considerable anxiety. Although it was still possible to import grains, primarily from the United States, prices rose dramatically owing to demand and to soaring shipping rates.[56] Domestic production had to increase; in January 1917, Agricultural Advisor Ole T. Bjanes proposed a major plan for cultivating more land and intensifying agricultural work. Each county and village would be mobilized, every farmer reached and impressed with the gravity of the situation. To support the local organizing efforts and the intensive planting, the Storting (Norway's parliament) granted 1. million crowns, primarily for low-interest loans for farm improvements and machinery. An additional 11 million crowns for 1917–18 was appropriated to subsidize the purchase of chemical fertilizer and feed concentrate. To coordinate local activities and administer funds, the Ministry of Agriculture established the Office for Production with Bjanes as director.[57]

Although the first effort to increase production, during summer 1917, proved a moderate success by increasing the total land cultivated by 12,5 percent, these gains were eclipsed by an American ban on exports. The United States entered the war in April 1917 and in July imposed a general export prohibition. She required neutral countries to negotiate new trade agreements, ostensibly to restrict further their commerce with the Central Powers. Norway's high-level delegation, lead by Nansen, received a favorable welcome in the

55. Hans Trøgstad, "Jordbruksproduktionens økning i krigsaarene," *Norsk landmandsblad: Ukeskrift for praktisk landbruk og meieribruk* (1919), 284.

56. *NOS XII 245*, Table 198, "Shipping Freight Rates," shows, for example, that grain sent from New York to England (Bristol) quadrupled in price from 1914 to 1915, and by 1918 the rate climbed to nearly twenty-five times the rate obtained in 1913; Table 195, "Gross Freight Earnings in Ocean Transport," provides indirect evidence of the staggering increases in shipping rates for all vessels: from 211,478 (× 1000 crowns) for 1914 to 1,107,177 (× 1000 crowns) for 1917.

57. Trøgstad, "Jordbruksproduktionens økning," p. 284–86.

United States, but prospects for a quick agreement and resumption of grain shipments appeared remote. Government calculations during fall 1917 revealed that Norway's food reserves would be exhausted in much less than a year.[58] This realization precipitated emergency actions.

Food rationing, resisted throughout the war, could no longer be avoided.[59] The government took control over all importation of grain and flour and became the sole legal purchaser of domestic grain.[60] Clearly the Ministry of Agriculture's experiment in increasing production would have to be attempted again, only more intensively. A one-million-*mål* increase in cultivated land (1 *mål* = 0.247 acres) over the 1917 efforts would be necessary to prevent a crisis. Responding to the ministry's proposals, the Storting appropriated twenty-seven million crowns to help increase production during the 1917–18 growing season. Other drastic measures included a law endowing the state with the power to compel farmers to cultivate unused areas, to conscript persons and horses for agricultural work, and to expropriate misused land.[61] As spring neared, all was set for a massive national undertaking to increase food production.

In February 1918, in the midst of the national discussions and legislative proceedings on the food crisis, the liberal newspaper *Tidens Tegn* published an article about a Swedish plan to provide weather forecasts to farmers by telephone.[62] The article noted that although nobody knew how effective this system might be in practice, reliable weather forecasts, even only a short period in advance, could prove significant. Claiming that meteorological science should be brought into the service of agriculture to the greatest possible extent, he regretted that in Norway, where a massive increase in food production had become a vital necessity, scarcely anything was being done in this regard. The article stated that when asked to comment on this development, Hesselberg (now director of the Norwegian Meteorological Institute in Christiania) played down the possibility of a comparable telephone-based weather service for Norwegian farmers.

Bjerknes responded to the article vigorously. First, writing to

58. Keilhau, *Norge og verdenskrigen*, pp. 209–13; Riste, *Neutral Ally*, pp. 197–212.

59. Keilhau, *Norge og verdenskrigen*, pp. 214–15, 265–68; Furre, *Norsk historie*, pp. 87–89.

60. Keilhau, *Norge og verdenskrigen*, pp. 239–240; Paul Borgedal, *Norges jordbruk i nyere tid* (Oslo, 1968), 3:180–81.

61. Trøgstad, "Jordbruksproduktionens økning," p. 286. Keilhau's work (*Norge og verdenskrigen*) provides complementary details for this discussion.

62. "Bedre veirvarsler for den økede matproduktion: Bønderne i Sverige skal hver dag faa utsigterne telefonert hjem," *Tidens Tegn*, 13 February. Sweden faced a comparable threat of famine.

Hesselberg, he described how both he and Helland-Hansen read the former's comments "with much distress," for in spite of the difficult conditions for issuing weather forecasts, "the situation is such that meteorology is obliged for the country's sake to do its utmost."[63] Furthermore, this urgent situation might well be the chance to get the funds meteorology ought to have. On the same day Bjerknes wrote to Holtfodt, saying that regardless of the fate of the proposed military field weather service, a new situation was at hand, one that required mobilizing all resources to prevent famine: "Under these conditions agriculture has the unconditional demand for all the support it can receive from meteorology's side." Bjerknes suggested an expansion of the earlier plan for a field weather service that would put agriculture's needs in the foreground. He argued for retaining military supervision, on the grounds that military morale would be strengthed by an active military role in food production; the weather service itself could then avoid civilian bureaucratic difficulties. Although Bjerknes had insisted that because of his earlier position in Germany, his name should not be publicly linked with the field weather service, now, "for his fatherland's welfare," he would do all he could—actively and publicly.[64]

As it turned out, rather than waiting for the military to establish a field weather service, Bjerknes and Hesselberg agreed to cooperate in a rapid expansion of weather services through existing institutions. In response to Bjerknes's harsh letter, Hesselberg revealed his private idea of expanding the existing summer forecasting service for eastern Norway and initiating summer forecasting—for West Norway from the Meteorological Observatory in Bergen and for North Norway from the Haldde Observatory.[65] Traditionally, forecasts providing general weather information for the next twenty-four hours were issued for only a few farming regions in eastern Norway and for the

63. Bjerknes to Theodore Hesselberg, 18 February 1918, copy, VpVA.
64. Bjerknes to Holtfodt, 18 February 1918, copy, VpVA.
65. Hesselberg to Bjerknes, 20 February 1918, VpVA. The meteorological observatory in Bergen, much smaller than that in Oslo, was primarily responsible for issuing storm warnings along the coast during the fall and winter. See, in addition to NMIÅ, B. J. Birkeland, "Det Meteorologiske Observatorium i Bergen," in *Meteorologien i Norge i 50 aar* (Christiania, 1917), pp. 112–18. Haldde Observatory in northern Norway was established primarily for geophysical research, which included the meteorology of the area. Issuing storm warnings for some areas on the northern coast had been attempted. Plans were under way at this time to develop a weather service for North Norway, especially to aid the fishermen on whose work the economy of the north largely depended. O. Krogness and O. Devik, "Halddeobservatoriet og dets meteorolgiske arbeider," in ibid., pp. 124–38; *Vervarslinga for Nord-Norge 25 år: Festskrift utgitt i anledning av 25-års jubileet 1. februar 1945* (Tromsø, 1945), pp. 7–28; O. Devik, *Blant fiskere, forskere og andre folk* (Oslo, 1971), pp. 50–71, 76–95.

Christiania vicinity.[66] Newspapers, placards on select buildings, lanterns hanging on the back of trains, and various other signals informed the inquisitive whether the next day would be wet, dry, or changeable. Of course such forecasts reached only a few people, and predictions of such generality, moreover, were little use in agriculture where weather conditions often change rapidly and vary over short distances. Hesselberg's private plan depended more on Norway's rapidly growing telegraph and telephone networks for spreading the forecasts effectively than on major changes in forecasting methods.[67] It was the wartime blackout in the exchange of international weather data, which made forecasting itself even more difficult, that had made Hesselberg cautious, especially in the newspaper interview.

Bjerknes understood from his knowledge of the German field weather services that a mere expansion of the preexisting forecasting methods would not fulfill the predictive challenges confronting them. Time was short; spring planting would begin in a matter of weeks. He and his assistants started immediately in February to consider appropriate forecasting methods and to mobilize local telegraph and telephone systems to spread predictions to farmers.[68] Bjerknes wanted some evidence for the possibility of positive results before turning to the government for assistance. Even while he and his Carnegie assistants, Jacob Bjerknes and Halvor Solberg, were experimenting and deliberating strategy, another opportunity enhanced the proposed weather forecasting's worth.

In Plans for Commercial Aviation

European aviation enthusiasts and entrepreneurs began to plan and establish commercial airline companies even before the war ended. In Norway, a group of prominent Norwegians, acting on expectations of postwar aviation, established in April 1918 the Norwegian Aerial Transport Corporation, (Det Norske Luftfartrederi A/S).[69]

66. See NMIÅ for the years before 1917. A general description of forecasting practices ca. 1916 is provided by Aage Graarud's contribution, "Veirvarselsavdelingen," in *Meteorologien i Norge i 50 aar*, pp. 67–85.

67. Hesselberg to KUD, 20 April 1918, KUD/D, folder: "Værvarslinga på Vestlandet," from file "Værvarslingstjenesten (eldre dokumenter) 1918–23" (hereafter cited as "VpV"/"Værtj.").

68. Bjerknes, "Hvordan Bergenskolen ble til," pp. 12–13; various letters exchanged in February and March by Bjerknes, Helland-Hansen, and B. J. Birkeland (of the Meteorological Observatory in Bergen) with the administrative leaders of neighboring counties and of telegraph offices show the preliminary organizing attempts and the enthusiastic support given by local officials. VpVA and BHH.

69. "Den norske luftfart," *Nordmannsforbundet* 11 (1918), 172–79; "Commercial Aviation in Norway," *Scientific American* 27 April 1918, 396; "A Norwegian Air Service," *Aeronautics* 14 (1918), 357. Both Kåre Fasting, *Fra Kontraskjæret til Tokio: Norsk Sivilflyg-*

With support from several major newspapers and from such famous persons as former prime minister Christian Michelsen and Fridtjof Nansen, the corporation sold shares of stock with relatively little trouble. Initial plans called for aviation to unify the nation and bring foreign markets closer. Travel time between the few major cities would be revolutionized; trips that took days and even weeks would take only hours by airplane. The entire enterprise took on a patriotic, almost romantic, quality. Recognizing the difficulties of flying over Norway's mountainous terrain, the planners envisioned most initial routes following the coast. Sea planes could serve the coastal routes as well as the inland ones because, to save money, the latter were to follow rivers and lakes, making expensive landing strips mostly unnecessary. Perhaps even more than internal air routes, the connections with Great Britain and the Continent raised the expectations of Norwegian businessmen. Early goals included regular airmail traffic between Aberdeen and Stavanger, which would connect by air Christiania, Christiansand, and Bergen in Norway and with major cities in Britain. Another route would connect with the Continent through Copenhagen.

In the earliest discussions on commercial aviation in Norway, meteorological needs were prominent.[70] To begin preparing weather services for the upcoming commercial routes, Norske Luftfartrederi's administrative director, Wilhelm Keilhau, and its technical director, Navy Capt. Gyth Dehli, went to Bergen in April and met with Bjerknes.[71] Bjerknes then composed a preliminary report on the measures needed to secure the weather services for commercial aviation.[72]

Bjerknes admitted in his report that details could only be worked out once experience was gained from both flying and the proposed summer weather service. Even this early, Bjerknes could note some necessities for a commercial aviation weather service. One was full information of atmospheric conditions from the major landing places for every flight. Safety depended on knowledge of possible storms, rain, thundery weather, and fog; economical operation, on knowledge of wind direction and speed at several heights to permit a choice of flight plan. Another necessity was the immediate training of a corps of "scientific meteorologists," at least fourteen. Bjerknes sug-

---

ing gjennom 50 år (Oslo, 1959), pp. 138–88; and Lindtveit and Thoresen, *På vingene over Norge,* pp. 104–16, provide details of the entire period under consideration. Another smaller corporation, *Nordisk Luft Kraft,* was also established at this time.

70. Einar Sem-Jacobsen, "Lufttrafik, med specielt henblik paa vore forhold," *Aeroplanet* 1, no. 2 (1918), 38.

71. Reported in Gyth Dehli to Norske Luftfartrederi, 27 May 1918, copy, VBP.

72. Bjerknes to Keilhau, 1 May 1918, KUD/D, "VpV"/"Værtj."

gested that a good first step would be for the Storting to appropriate funds for seven trainees, who could then help in the upcoming proposed expansion of summer forecasting. Last, he noted the necessity of starting to build a network of wireless telegraphy stations around the country, to serve both aviation and weather forecasting, the national telegraph service being already overloaded to the bursting point. Clearly, Bjerknes more than ever viewed aviation as a means of obtaining resources.

Bjerknes's Proposal in the National Context: Self-Sufficiency and Science

In a proposal sent later in April to the Ministry for Church and Education, Bjerknes noted all the reasons "the significance of a really effective weather service can scarcely be exaggerated."[73] Looking beyond the immediate food crisis, he pointed out that before the war, agricultural production was valued at close to two million crowns. If the yield from agriculture could be increased by merely one percent through effective weather forecasting, then the price of the entire forecasting service would be many times paid. Furthermore, in addition to the direct increase in the yield from agricultural production, of no less importance would be the savings in labor-intensive periods during planting and hay making (*onnearbeidet*): reliable forecasts could help farmers plan their work. He also pointed out that aviation would soon make very great demands on the meteorological service. The proposed emergency summer forecasting should be just a trial aimed at finding the principles upon which to organize an entire meteorological service for the future.

To avoid delay and extra costs, he proposed that the Bergen Museum's Geophysics Institute work with the Bergen Meteorological Observatory, which before the war was responsible for issuing coastal storm warnings. Bjerknes's own department of the institute actually existed only on paper; the observatory at least provided a locale. In the meantime Hesselberg had also sent a proposal for expanding during the summer the existing weather service for agriculture in eastern Norway: "The significance of such forecasts will always be very great, but now, when the whole country's welfare depends upon our own food production, they are more important than ever. It is therefore now necessary to put forward a plan to obtain better weather forecasts for farmers."[74]

73. Bjerknes to KUD, 22 May 1918, KUD/D, "VpV"/"Værtj."
74. Hesselberg to KUD, 20 April 1918, KUD/D, "VpV"/"Værtj."

The Ministry of Church and Education supported the two proposals and requested from the Storting 22,300 crowns for the expanded service for eastern Norway and 100,000 crowns for the new
service for western Norway for July, August, and September 1918.[75]
Other ministries were asked their opinions on the proposals. Commerce recommended support, thinking of civil aviation's coming requirements. Defense was somewhat reserved; officers connected with
the air forces supported the proposal, while others thought the need
not yet so pressing. The Ministry of Agriculture also strongly recommended the appropriations, but one official in the ministry wrote a
memorandum claiming that a weather service for West Norway was
not all that significant because the important farming regions were in
the east.

With the aid of Sem Sæland, physicist and then member of the
Storting, Bjerknes met Prime Minister Gunnar Knudsen. After only
fifteen minutes of explaining how an expanded weather service might
benefit agriculture, Knudsen allegedly commented, "Yes, it is of
course obvious; this we must have."[76] Knudsen, who also held the
cabinet post of minister of agriculture, disagreed with the ministry's
official who opposed funding for West Norway. The Ministry of
Church and Education then discussed the issue with Knudsen and
Bjerknes and subsequently gave full support to the new weather
service.[77]

By 26 June 1918, when the Storting approved these appropriations, Bjerknes, Jacob, and Solberg had already set up an experimental weather service. When the summer forecasting season formally
began in July, Solberg went to Christiania to lead the expanded service for East Norway (*Østlandet*), while Jacob assumed forecasting responsibilities for the Bergen-based service. Bjerknes found himself
organizing and directing the operation. In time of national crisis, he
"suddenly became a 'practical' meteorologist,"[78] responding to a food
shortage. Some comments are in order on the broader national context of his and the government's response to the food crisis.

Although neutral, Norway underwent major social, economic, and
political changes during the war. The enormous profits from the
merchant marine and rampant speculation made Norway a creditor
nation for the first time in modern history. Fishery enjoyed great

75. *Forhandlinger i Stortinget 1918* (Proceedings of the Norwegian parliament), Storting proposition nos. 110 and 125.

76. Bjerknes, "Hvordan Bergenskolen ble til," p. 13.

77. *Forhandlinger i Stortinget 1918*, indst. s. no. 203, p. 345; "P. M. Veirvarslinger for
Vestlandet og Trøndelagen 30.5.1918" and note dated 4 June 1918, KUD/D, "VpV"/-
"Værtj."

78. Bjerknes to Arrhenius, 4 June 1918, SAP.

profits because the British attempted to limit sales to German markets by buying up much of the Norwegian catch. The new wealth, especially among shipowners and merchants, stood out in a traditionally poor and relatively homogeneous society. Fixed-wage earners and industrial workers suffered increasingly as food and housing shortages, coupled with the unprecedented inflationary surge, led to astronomical prices for basic commodities. Social unrest ensued, particularly after 1916.[79] Inflation ran rampant; shortages in industrial raw materials and food stuffs began to have serious effects on economic performance and social stability.

Increasingly, Prime Minister Knudsen's liberal (Venstre) party sought to alleviate these problems by expanding the state's role in commerce and social welfare. At times acting for the benefit of private enterprise and at times acting as a business in itself, Knudsen's government both created new ministries and enlarged existing ones for obtaining and managing raw materials.[80] Of particular importance were the efforts to find Norwegian resources to replace materials normally imported. The year 1917 witnessed a growing wave of enthusiasm for self-sufficiency (*selvforsyningsbegeistring*).[81] Not only was the duration of the war uncertain—and with it, imports of building materials, fuels, ores, and food stuffs—but some politicians and industrialists also believed that extreme international economic competition after the war would require a greater degree of national self-sufficiency even then.

Following the example of scientists in other nations, some Norwegian researchers recognized the opportunity to improve the conditions for pursuing their research while also contributing to their nation's welfare. There were several attempts to bring together government, private industry, and scientific institutions to help Norway achieve greater self-sufficiency. The president of the Norwegian Academy of Science and Letters, W. C. Brøgger, a professor of geology, called for the establishment of a national research council, like that recently established in the United States through the efforts of G. E. Hale and Robert Millikan.[82] Brøgger and others pointed to the Ameri-

79. Vogt's *Jerntid og jobbetid* provides factual and impressionistic accounts of all aspects of Norwegian life during the war.

80. T. C. Wyller, "Utvidelsen av statens myndighetsområde i Norge under første verdenskrig," *Historisk tidsskrift* 39 (1959), 321–37; K. Tønnesson, *Sentraladministrasjonens historie: IV. 1914–1940* (Oslo, 1979), pp. 1–46; Vogt, *Jerntid og jobbetid*, pp. 110–111.

81. Vogt, *Jerntid og jobbetid*, p. 120.

82. W. C. Brøgger, "The National Research Council, dets grundlægelse og maal: Foredrag i fællesmøtet 1ste december 1916," *Oversigt over Videnskabsselskabets møter i 1916* (Christiania, 1917), pp. 44–54.

can example as proof that mobilizing, coordinating, and strengthening scientific research efforts were as essential to the neutrals as to the warring nations.

In 1917 the new Ministry for Industrial Supplies established a Raw Materials Laboratory at the Mineralogical Institute at the university in Christiania, where Prof. V. M. Goldschmidt and his staff began testing Norwegian minerals for substitutes to imported materials. Furthermore, the Central Committee for Scientific Cooperation for the Promotion of Industry (Council for Applied Science, from 1921 until its decline in 1926) was created to bring together representatives of major scientific institutions and of industry. It aimed both to engage existing institutions in assisting particular industrial sectors and to draw up plans to establish a series of government-sponsored industrial research laboratories; greater national self-sufficiency was the goal. Newspapers devoted considerable attention to these first steps toward setting Norway's economy on a "scientific foundation."[83] Science, moreover, began to assist in rationalizing sectors of society in which the state had begun to be engaged, agriculture for one.

83. V. M. Goldschmidt, "Om Raastofkomiteens arbeide og resultater," *Teknisk Ukeblad* 37 (1919), pp. 6–10; "Norske raastofe: Videnskab og industri samarbeider," *Aftenposten,* 30 October 1918; "Naar Videnskab og Industri samarbeider," *Verdens Gang,* 7 March 1918. A comprehensive history of the various attempts to mobilize science during the war can be found in John Peter Collett, "Videnskap og politikk: Samarbeide og konflikt om forskning for industriformål, 1917–1930" (Diss. *hovedoppgave,* Dept. of History, Oslo Univ., 1983).

[6]

An Experiment in Forecasting
Leads to a New Cyclone Model

B JERKNES'S primary goal for the summer of 1918, and the
government's hope, was to use weather forecasting to increase agricultural production. At the same time, Bjerknes sought to
experiment with possible forecasting practices that could meet the
coming needs of aviation. The challenge was to achieve precision and
reliability on a scale hitherto never attempted, and conditions for
issuing precision forecasts could scarcely have been worse.

Before the war, forecasting in Norway relied heavily on weather
data, sent by telegram, from Britain, Iceland, and the Faroe Islands.
Now, in time of war, German Zeppelin raids on Britain, and forecasting in most military operations, these data had become secret. Weather
predictions for Christiania and eastern Norway suffered in accuracy,
and the storm warning service for the west coast had been suspended
outright.[1] What is more, other than these general storm warnings for
the coastal regions, organized weather forecasting had never been
tried in West Norway and Trøndelag. This area's varied and mostly
rugged terrain and its frequent storms and rapid weather changes
posed immense problems for a weather prediction service. Without
data from the North Sea and beyond to locate approaching low-
pressure systems (cyclones), traditional methods based on delineating
the field of surface pressure were thoroughly useless. Of course
Bjerknes rejected on principle these formalistic "statistical-climatological" methods. In establishing a forecasting service, he wished

1. NMIÅ *1916–1917*, p. 8; Theodore Hesselberg to V. Bjerknes, 9 February 1916,
JBP.

to use and extend the work on rational predictions based on a dynamic comprehension of the atmosphere. Bjerknes intended to model this system structurally after military field weather services and to focus attention on lines of convergence in the wind field. Special resources available during the war offered hope that the motion of such lines, and their related rain areas, could be calculated as they approached the coast.

Bjerknes knew that field weather services at the battle fronts had shown that to compensate for weather data unavailable from enemy territory and to attempt detailed, precise, short-term forecasting for specific locations, it was necessary to establish a considerably denser network of observation stations than those used in traditional weather forecasting. Similarly, to compensate for the stoppage of weather telegrams from across the North Sea and to achieve geographic and temporal precision in the agricultural and the contemplated aviation forecasts, an increase in the number of observation stations was essential.[2]

Having produced the crises leading to the establishment of a new Norwegian weather service, the war ironically also furnished resources that might be used to overcome the barriers to successful forecasting. Principal among these were the navy's U-boat watch stations. In its effort to protect Norway's neutral waters from submarine incursions, the Norwegian navy established a network of observation stations on the islands and skerries comprising Norway's North Sea archipelago.[3] Located on the western-most outcroppings of land facing the North Sea, and manned by sailors and fishermen who were experienced weather observers, these installations could also serve as lookout posts for approaching weather systems.[4] With this in mind, Bjerknes contacted the chief of the navy's Bergen district, Com-

2. Although Bjerknes previously had stated that ideally the space-time parameters for observations should approach differential changes, his primary inspiration here came from the field weather services. His comments to R. S. Woodward (16 August 1917, copy, VBP) urging the establishment of an American aerological observation network show that in general his notion of what would be an adequate network was considerably more sparse than that used in field weather services. Similarly, Sverdrup at first opposed the idea of a dense network of surface stations, claiming that as long as the upper-air observations were so sparse, little benefit would come from increasing the number of surface stations. (Tor Bergeron claims that Solberg told him this story when they first met in fall 1918 (lecture, 17 September 1953, TBP.)

3. On the diplomacy and economics of maintaining neutrality, see Olav Riste, *The Neutral Ally: Norway's Relations with Belligerent Powers in the First World War* (Oslo, 1965), pp. 123–76; Wilhelm Keilhau, *Norge og verdenskrigen*, Carnegie Endowment for International Peace: Economic and Social History of the War, Scandinavian Series (Oslo, 1927), pp. 59–63, 138–52, 178–208.

4. Finn Pedersen, "Vervarslinga på Vestlandet, 1918–1968," in *Dette er Vervarslinga på Vestlandet . . . : Utgitt til 50–årsdagen 1. juli 1968* (Bergen, 1968), p. 15.

modore Ragnar Rosenquist, and met with him on 21 February 1918. Having emphasized the potential benefits to domestic food production that an accurate weather forecasting service could offer, Bjerknes quickly obtained Rosenquist's assistance.[5] Not only were the observation stations placed at Bjerknes's disposal; so too were naval vessels for transporting him to various remote locations off the coast. During spring 1918, Bjerknes received pledges for cooperation from the chiefs of the other navy districts on the Norwegian coast, from the North Sea off Trøndelag to the Skagerak. Using these stations and their personnel, Bjerknes and his assistants began studying the feasibility of erecting a weather service well in advance of approaching the government.

Starting with the then-existing nine telegraph stations in Norway, of which only three were on the west coast, Bjerknes expanded this network. Ten navy observation stations were incorporated by the end of February, and by the end of June, just before the start of the official forecasting effort, he had set up sixty new weather stations in West Norway (Fig. 3).[6] With the help of the navy vessels at his disposal, Bjerknes had access to remote and scattered locations, thereby making possible a further strengthening of the observation network by including a number of lighthouse keepers, farmers, and fishermen.

The navy lookout stations offered a unique opportunity for meteorologists to obtain precise wind observations. Traditional weather forecasting techniques relied little on wind measurements, nor could fine differentiations in the wind's direction easily be measured. Observers generally recorded wind direction in only one of eight, or at most sixteen, compass points (e.g., NNW).[7] In contrast, the Norwegian navy observers, who were equipped and trained to locate and to pinpoint submarines that entered their territorial waters, could apply their skills to measure the wind with great accuracy. Bjerknes found that they could specify wind direction to within 5-degree increments. Furthermore, wind measured at these exposed islands and reefs would reflect truer atmospheric conditions than wind observations taken over land, which are subject to distortion from obstacles and friction. This unprecedented opportunity to obtain precise wind measurements from a dense network of stations offered hope for

5. V. Bjerknes, "Hvordan Bergenskolen ble til," in *Vervarslinga på Vestlandet 25 år: Festskrift utgitt i anledning av 25-års jubileet . 1. juli 1943* (Bergen, 1944), pp. 12–14; see also correspondence in VpVA and BHH.

6. NMIA *1917–1918*, p. 13.

7. International Meteorological Committee, *Report of the Tenth Meeting, Rome 1913* (London, 1914), p. 9.

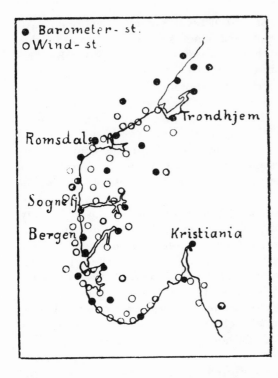

Fig. 3. Observation stations for summer 1918 forecasting. From V. Bjerknes, "Veirforutsigelse: Foredrag ved Geofysikermøtet i Gøteborg, 28. august 1918," *Naturen* 43 (1919), p. 7.

precalculations of the wind field. It also offered a potential new means of predicting rain: charting and precalculating the movement of lines of convergence in the wind field.[8]

Lines of Convergence, Summer Forecasting, and Discovery

In Leipzig, Jacob Bjerknes had investigated the frequency with which lines of convergence passed over Europe and the relation of these lines to line squalls and to other meteorological phenomena. His principal focus then was a kinematic analysis of the wind field in which the lines of convergence occurred.[9] The work fit into the general efforts of Bjerknes's Leipzig school toward rational precalculations using kinematics and dynamics. Concurrent work included Vilhelm's, as well as Sverdrup's and Holtsmark's, attempts at dynamic predictions of the wind field.[10] Using the methods for graphic differ-

8. V. Bjerknes, "Veirforutsigelse: Foredrag ved Geofysikermøtet i Gøteborg, 28. august 1918," *Naturen* 43 (1919), 8–9.

9. J. Bjerknes, "Ueber die Fortbewegung der Konvergenz- und Divergenzlinien," *MZ* 34 (1917), 345–49.

10. V. Bjerknes, "Ein prognostisches Prinzip der dynamischen Meteorologie," *Berichte SGW* 69 (1917), 278–86; H. U. Sverdrup and J. Holtsmark, "Ueber die Rei-

entiation developed in *Dynamic Meteorology and Hydrography,* Jacob analyzed the kinematics of the wind field near lines of convergence. He found that in the immediate vicinity of the lines, the wind field exhibits a distinct pattern of vortices. By applying the theorems for the formation of vortices arising from the lines of flow in the wind field, he arrived at an equation expressing the propagation of the lines of convergence as a function of the vorticity gradient. In addition to this quantitative relation, he noted a simple rule for predicting their direction: in the northern hemisphere, when looking along a line, in the direction of the wind converging toward it, the line of convergence will tend to move toward the right. When he compared the charts of wind flow with charts of weather distribution, Jacob noted two characteristic rain patterns associated with lines of convergence: one, a wide band-shaped rain area; the other, a narrow line squall. Although this relationship called for further investigation, it appeared that a rational means of predicting the movement of lines of convergence in the wind field could perhaps be useful in predicting line squalls and, more generally, rain.[11] Qualitatively, a relationship between lines of convergence and the other weather phenomena had long been recognized; the innovation and the promise of Jacob's short study was the introduction of rational methods for analysis and prediction.[12]

When in February 1918 the Bergen group came to consider means of predicting weather for West Norway, these lines and the prognostic equations appeared to provide one possibility for forecasting rain. If lines of convergence in the wind field could be identified as they approached the observation posts off the coast, then their movement and speed, along with that of their accompanying rain areas, could perhaps be calculated, making it possible to issue reliable predictions of rain to farmers at least several hours in advance of the rain.[13] By focusing on these discrete structures, they hoped to specify as well the distribution of rain to a greater extent than possible with diffusely defined low-pressure systems. This method would depend not on inaccessible knowledge of the weather beyond the North Sea horizon but on the precise wind measurements available from the coastal observation stations.

---

bung an der Erdoberfläche und die direkte Vorausberechnung des Windes mit Hilfe der hydrodynamischen Bewegungsgleichungen," *VGL,* vol. 2, bk. 2 (1917); V. Bjerknes, "Jahresbericht 1915–1916," draft of report to Carnegie Institution, Carnegie File, VBP.

11. J. Bjerknes, "Ueber die Fortbewegung," pp. 345–46.

12. Lines of convergence had been noted to have a relationship with line squalls and with cold and warm waves in *Dynamic Meteorology and Hydrography,* 2:60.

13. V. Bjerknes, "Veirforutsigelse: Foredrag," pp. 5–8.

Even before the group initiated trial forecasts, Jacob had resumed his earlier work on lines of convergence. In a preparatory investigation designed "to test the possibility of better and more detailed forecasting for short-time periods," Jacob gained new insight into the structure of cyclones. He collected observations previously recorded at Scandinavian climatological stations (observation posts that do not send data via telegraph to central weather bureaus and therefore numerous) to try calculating the actual movement of lines of convergence.[14] From this study, he found that such rational predictions were not so easily accomplished; he also happened upon an intimate connection between lines of convergence and cyclones. Rather than being characterized by a symmetrical spiral in the wind field, a cyclone seemed to consist of two differing air currents—one cold and one warm—that meet in a line of convergence running through the cyclone.[15] Once forecasting trials were underway Jacob gained further significant insights that led him to refine this picture and arrive at an innovative cyclone model.

Jacob's diary from the summer forecasting reveals his increasing familiarity with how lines of convergence related to cyclones.[16] The weather service provided forecasts for individual districts; southern Norway was divided into nineteen, of which seven fell under Bergen's jurisdiction. Observers along the coast and inland collected data at 8:00 AM and sent these by express telegraph to Bergen. By 9:30–10:00 AM, Jacob relayed the appropriate predictions to telephone stations scattered throughout the countryside, from which farmers could receive information. These predictions were only valid until 8:00 PM the same day; lack of foreign observations and prior experience in forecasting for West Norway mandated caution. An additional set of observations from 1:00 or 2:00 PM served as a control on the forecasts. (In Christiania, where the generally eastward march of weather systems could be followed more reliably, forecasts were issued in the afternoon, valid from 7:00 PM the same day until 7:00 PM the next.) Each forecast was actually an investigation; candid expressions such as "probably," "possibly," "I guess," and "in all likelihood" appear often in Jacob's diary. Because he could not yet rely on the calculation of the movement of lines of convergence, Jacob used a

14. See letters to Nils Ekholm in which V. Bjerknes requested data for this study, 2, 11 January 1918, Ekholm Papers, KVA-SUB. Reference to the plan to calculate the motion of lines of convergence is also mentioned in V. Bjerknes, "Veirforutsigelse," pp. 5–8.

15. J. Bjerknes, Untitled MS, "I forbindelse med nogen cyklonundersøkelser . . . ," n.d. [Spring 1918; probably lecture held March 1918], Box I(1), JBP.

16. "Veiroversigter for Vestlandet sommeren 1918 I" and "Daglige veiroversigter sommeren 1918 II," JBP.

variety of criteria in forecasting; the lines of convergence were never-theless central.[17] By controlling his morning predictions against ob-servations taken later in the day and by attempting to understand the changing weather patterns, Jacob learned more about lines of convergence.

From his preliminary investigations during the winter and spring, Jacob had assumed that the lines of convergence associated with dif-ferent rain patterns arose from different physical processes.[18] Those associated with broad rain patterns seemed to have warm air flowing into the back side of the lines; alternatively, cold air flowed in from the rear of the lines of convergence associated with narrow rain bands. Sometimes the line of convergence entered the cyclone's cen-ter in the same direction that the cyclone was moving. In July he identified the "warm" line of convergence as the one that points the direction of a cyclone's movement.[19] Increasingly, he suspected that a cyclone actually possessed two lines of convergence.

A situation on 14 August posed a special problem. All evidence in the morning indicated a cyclone to the northwest of West Norway, as yet off the map, that later in the day would move southward, parallel to the coast: therefore it should rain along the coast but not on Trøndelag or inland.[20] Later in the day, a line of convergence was identified along the Sunnmøre-Romsdal coast (Fig. 4). Warmer air flowing into the line from behind suggested the line was of the variety that points the cyclone's direction; only in this case, it indicated that the cyclone would move in an easterly direction. All other prognostic indicators, including a rapid fall of barometric pressure along the coast, suggested a southerly course. In the diary for 15 August, Jacob wrote, "All the same, the cyclone didn't go southward"; it followed the line eastward. From then on, he called this line of convergence the "steering line." Jacob had also noticed that on the morning map of 15 August the other type of line of convergence had crossed the west coast along a north–south axis.[21] The drop in pressure along the coast the previous day was actually connected with this line of con-vergence and not with the cyclone per se. Although the center of the cyclone was still at sea off the map's boundaries, it would seem that the two lines must meet at its center. Jacob was convinced that rather

17. Ibid.

18. J. Bjerknes, "I forbindelse med nogen cyklon. . . ."

19. J. Bjerknes, "Veiroversigter I," entry for 10 July.

20. Ibid. II, entry for 14 August; V. Bjerknes, "Veirforutsigelse," *Skandinaviska geo-fysikermötet i Göteborg den 28–31 augusti 1918*, ed. O. Nordenskjöld and H. Pettersson (Göteborg, 1919), pp. 6–8.

21. J. Bjerknes, "Veiroversigter II," 15 August.

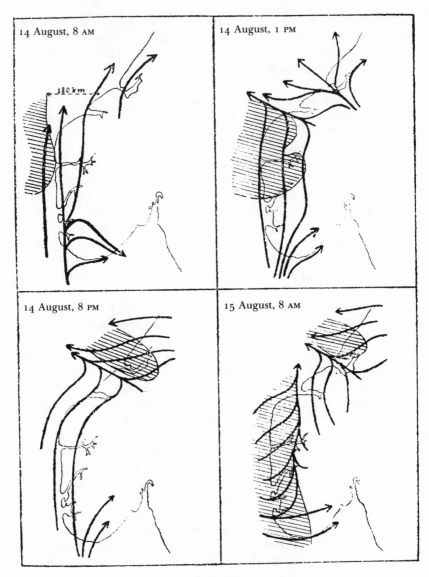

Fig. 4. Lines of convergence, 14–15 August 1918 (redrawn). From V. Bjerknes, "Veirforutsigelse," p. 11.

than just being loosely associated with cyclones, the twin lines of convergence were a fundamental characteristic.

In Jacob's initial model, the two lines of convergence enter to the right of the cyclone's path and meet at its center (Fig. 5). The convergence line in the rear of the cyclone comes in almost perpendicular

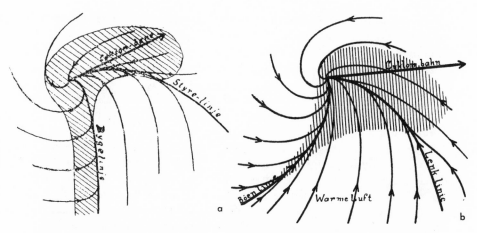

Fig. 5. Early depictions of the cyclone, summer and fall 1918. Vilhelm Bjerknes presented this very first drawing (a) at a geophysicists' meeting at the end of August. Styre-linje=steering line; Bygelinje = squall line; Cyklon-bane = path of cyclone. From V. Bjerknes, "Veirforutsigelse," *Skandinaviska geofysikermötet i Göteborg den 28–31 augusti 1918* (Göteborg, 1919), p. 5. In versions of this lecture that he sent in the early fall for publication in scientific journals, the cyclone model (b) has been modified. Although rain is still depicted in the warm sector, the pattern of wind flowing into the lines of convergence is more stylized, resembling the kinematic diagrams in *Dynamic Meteorology and Hydrography*, vol. 2 (see Fig. 8 below). Lenklinie = steering line; Böenlinie = squall line; Warme Luft = warm air. From V. Bjerknes, "Wettervorhersage," *Meddelanden från K. Vetenskapsakademiens Nobelinstitut* 5 (1919), 4. This version is actually Jacob's intermediary model (compare Fig. 10b).

to the path of the storm and is identified as the line squall, which had traditionally been thought to follow behind—and not actually to be part of—a cyclone. The other line increasingly shares the same direction as the storm where it enters the cyclone's center. A line drawn tangent to this steering line at the point where it meets the squall line at the center, reveals the instantaneous direction of the cyclone, thereby providing a means of predicting a cyclone's movement.[22]

Additional secrets about the cyclone's nature were revealed in the charts based on data telegraphed from the dense network in West Norway. The air between the two lines of convergence of a cyclone was warmer than that bordering on the outside of the lines. Traditionally, a cyclone was usually thought to be a relatively symmetrical structure with a central core of warm or cold air; the new model

22. Reported in V. Bjerknes, "Veirforutsigelse: Foredrag," pp. 2–3.

seemed to possess an asymmetrical thermal structure consisting of a distinct "tongue" of warm air bounded by colder air. In a similar manner the rain pattern in a cyclone, which generally had been thought to be distributed rather symmetrically around the cyclone's central core, appeared to be connected with the lines of convergence composing the cyclone. Rather than being a function of the distance from the cyclone's center, rainfall seemed to be linked with the location of these lines.

These discoveries were immediately put to use in the forecasting service. Using the tangent to the steering line to predict the motion of cyclones and their lines of convergence, Jacob Bjerknes lowered the percentage of erroneous forecasts.[23] By the end of the summer forecasting season, Vilhelm and Jacob did not as yet know how successful their experiment had been in assisting the farmers in West Norway. They did recognize, however, that this unique effort based on an unprecedented dense observation network and on exceedingly precise wind observations yielded intriguing insights that required further examination.

A Preliminary New Cyclone Model

In October 1918, Jacob Bjerknes wrote his now-classic paper "On the Structure of Moving Cyclones," in which he attempts to explain the structures found on the summer weather maps.[24] A common practice in the historical literature on this subject has been to read into this paper concepts and ideas that only later emerged. It has often been incorrectly claimed that the notion of a weather "front" and a cyclone model based on fronts arose from the summer 1918 work. The Bergen concept of a front was clearly articulated first in 1919. Two-dimensional geometric lines of convergence in the wind field are not the same as three-dimensional weather fronts. A concept of fronts did not simply spring from the summer data. Fronts were not discoverable in the sense of a phenomenon waiting out in nature to be seen; they were an interpretation. How Jacob regarded his data and how he interpreted them were conditioned by his past experiences. He had been trained in the analytical methods and conceptual schemes of the Leipzig school; his initial cyclone model was an innovation within that explanatory framework. He had yet to alter that framework. He tried in fact to build his cyclone model chiefly by using older kinematic depictions of atmospheric phenomena. His ini-

23. J. Bjerknes to Svante Arrhenius, 25 October 1918, SAP.

24. J. Bjerknes, "On the Structure of Moving Cyclones," *GP* 1, no. 2 (1919); also in *MWR* 47 (February 1919), 95–99. Unless otherwise specified, citations are from *MWR*.

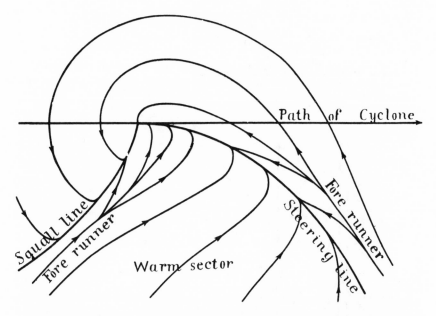

Fig. 6. Lines of flow, J. Bjerknes's cyclone model, October 1918. Note a "fore-runner" (line of divergence) is associated with the steering and squall lines. From J. Bjerknes, "On the Structure of Moving Cyclones," *GP* 1, no. 2 (1919), 2.

tial cyclone model and those that followed, owed as much, if not more, to interpretive construction as to empirical discovery.

Jacob Bjerknes organized much of his article around an idealized cyclone model defined by the flow of air (Fig. 6). Here, on a horizontal projection, such as one might find on a weather map showing the atmospheric conditions at ground level over some geographic area, he depicted a cyclone consisting of two lines of convergence into which warm and cold air are flowing: the steering and squall lines. Unlike the original schematic diagram from August, the lines of flow are more complex, also revealing so-called forerunners, or lines of divergence in advance of the lines of convergence. We recognize as well from the earlier depiction a "warm sector" between the two lines of convergence and the tangent line at the cyclone's apex, which shows the direction of the storm at that moment. This portrayal of a cyclone's horizontal features at the surface was supplemented by vertical cross-sections showing the flow of air above the steering and squall lines. Jacob organized the model around the specific pattern in the horizontal lines of flow consisting of a line of divergence leading a line of convergence. This configuration structured the development

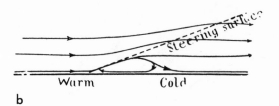

b

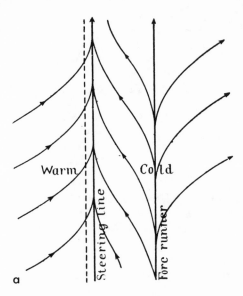

a

Fig. 7. Steering line and surface: (*a*) horizontal field of flow, (*b*) motion in vertical projection. From J. Bjerknes, "On the Structure of Moving Cyclones," p. 2.

of the models because he devised vertical models that could account for the appearance of these parallel lines.

At the steering line (Fig. 7a), Jacob assumed that, at ground level, warm and cold air respectively flow in toward the line from both sides but do not mix. Rather, they flow upward along a surface boundary, or "steering surface" (Fig. 7b), that separates them. The cold air, being dense, can only rise some distance before it sinks back to the surface, where it then spreads out, forming a line of divergence at the ground. The "rolling" wedge of cold air thus formed diverges at that surface and moves off horizontally, perpendicular to the line of divergence, thereby allowing the entire system to propagate forward.[25] Furthermore, he maintained that a line of divergence and its attendant "rolling mass" of cold air must exist in order for a steering line to

25. Ibid., p. 96.

propagate forward: "The existence of the forerunning line of divergence seems thus to be a necessary condition for the propagation of the steering line."[26] Why the emphasis on the line of divergence? In fact, Jacob himself later abandoned this practice of delineating paired lines of convergence and divergence. Furthermore, the depiction of a line of convergence that he used in this article, in which lines of flow enter from both sides toward a central line, he later disregarded also, thinking it "a rather special one [which] does not often occur in nature."[27]

Jacob's initial depiction of a cyclone originated from his use of existing kinematic models of atmospheric motions. In *Dynamic Meteorology and Hydrography*, volume 2, lines of convergence and divergence on the one hand, and cyclones on the other, are analyzed in terms of the kinematics of wind flow.[28] Jacob attempted to alter the 1911 kinematic model of a cyclone (Fig. 8), with its wind spiraling inward to a "convergence point," by integrating convergence and divergence lines into the spiraling motion. He claimed that once the number of observations became dense enough, it was possible to discover "several deviations from the regular spiralic shape" of the cyclone, namely the linked lines of convergence and divergence and the asymmetrical thermal properties.[29] The idea that forward propagation can occur only when lines of convergence and divergence are linked together, came also directly from the older work's claim that "a rolling mass of air will be bounded at the surface by a line of convergence and a line of divergence parallel to each other."[30] Not only were the linked lines of convergence and divergence taken from the 1911 work; Jacob portrayed the vertical cross-section of the steering line from the earlier kinematic models, but with the important addition of inserting a boundary surface separating the two types of air (Fig. 9). Indeed, the confusing diagrams and descriptive accounts in which Jacob portrays air flowing through the surface boundaries seem to imply that such boundary surfaces were introduced more on an ad hoc basis than through rigorous conceptualization. Having previously learned of Max Margules's theoretical calculations showing that a discontinuity can form and be maintained when two differing

26. Ibid., pp. 96–97.

27. J. Bjerknes and H. Solberg, "Meteorological Conditions for the Formation of Rain," *GP* 2, no. 3 (1921), 9.

28. Compare with that volume's pp. 48–55 and V. Bjerknes, "Synoptical Representation of Atmospheric Motions," *QJRMS* 37 (1910), 267–81.

29. J. Bjerknes, "Structure of Moving cyclones," p. 95.

30. *Dynamic Meteorology and Hydrography*, 2:53–54.

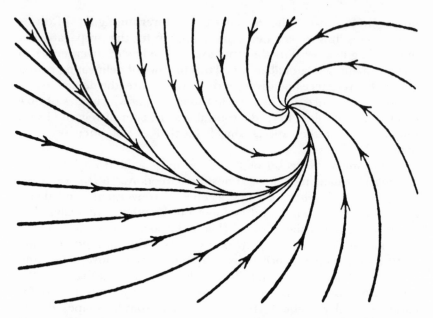

Fig. 8. Lines of flow: cyclone and line of convergence (1911 depiction); one of several "complex singularities" in the two-dimensional vector field of motion. The pure spiral form of a cyclone, identified with a convergence point, is here modified slightly to include a line of convergence. From *Dynamic Meteorology and Hydrography* 2:58.

air currents meet, Jacob incorporated this postulate into the preexisting kinematic models.[31]

31. The theoretical possibility of the existence of a discontinuity between two different air currents was well known at this time, having been postulated by Helmholtz, and more elaborately by the Austrian meteorologist Max Margules. Their work was discussed, along with other recent literature, at the Leipzig institute seminar; but it would be incorrect to interpret Jacob's work as an elaboration of these theoretical postulates and computations. Margules's work was an intellectual precondition but not a causal "influence"; it provided Jacob with a precedent for conceiving a discontinuity between two air currents and offered an equation for calculating the slope of these boundary surfaces based on the temperature and wind velocity on either side of the discontinuity. For Jacob, remember, lines of convergence provided the starting point for the cyclone model, and Sandström had delineated them independent of any thoughts of atmospheric discontinuities or the source of energy in cyclones. When Margules's articles were read in Leipzig, no one thought of discontinuities, of connections to lines of convergence and line squalls; Margules's work was considered mainly with regard to atmospheric thermodynamics and energy balance. Vilhelm's lecture notes (VBP) from Leipzig show clearly that even after reading Margules's work, cyclones were still viewed as spirals in the wind field. In claiming that Margules's work neither caused or led to conceptual change, I am not denigrating it. Austrian and German meteorologists who later claimed the Bergen meteorology offered little new or failed to credit "precursors"

Fig. 9. Lines of flow: rolling mass of air bordered by a line of convergence and a line of divergence; one of several "complex singularities" in the two-dimensional vector field of motion. The top part of the diagram represents a verticle cross-section above the lines of convergence and divergence. From *Dynamic Meteorology and Hydrography* 2:59.

There are similar conceptual difficulties with the squall line. Jacob first modeled the vertical cross-section on Wilhelm Schmidt's well-known laboratory model of a line squall. When he realized this experimentally derived model did not permit the inclusion of a line of divergence linked with and leading the squall line, Jacob soon discarded it.[32] In a report to the government in early December, Jacob

often appropriated Margules for their cause, pointing to his tragic death by starvation in postwar Austria, his eccentric nature, and his early retirement from the Austro-Hungarian meteorological service. His lack of impact and his professional frustration need more analysis. True, his works were written in a heavy style; yet another fact is overlooked: Margules was a Galician Jew, who twice refused to convert to Christianity and therefore twice had his scientific career in Vienna thwarted.

32. J. Bjerknes, "Structure of Moving Cyclones," p. 97. Again Jacob assumed that warm and cold air flow toward one another and meet, but do not mix, at the squall line. The "squall surface" with a characteristic "head" at its front, which was introduced as the boundary surface between the warm and the cold air currents, was modeled after Wilhelm Schmidt's experimental hydrodynamic model of a line squall; his "Zur Mechanik der Böen," *MZ* 28 (1911), 355–62; esp. 356–57. In response to V. Bjerknes's request for information on squalls and thunderstorms, Alfred Wegener pointed to Schmidt's work in a letter of 4 May 1915, VBP. Jacob discarded this initial depiction

attempted to achieve a vertical structure that could account for a line of divergence accompanying the squall line.[33] This depiction, never published and certainly discarded very quickly, showed warm air sinking at the squall line, revealed air flowing through the "squall surface" boundary that was supposed to separate the cold and warm air, and provided no mechanism to account for the squall line's showers and thunderstorms.

Jacob's presentation of the weather phenomena associated with a cyclone further demonstrates that this cyclone model was developed ad hoc, grounded in the kinematic analyses, and built up piecemeal using preexisting theory and models.[34] In the first illustrations, a rain and cloud pattern was overlaid upon the lines of flow and lines of convergence. Rough drafts of these illustrations show that the pattern used was taken with slight modification from Julius Hann's classic textbook description of weather in a cyclone (Fig. 10).[35] Hann's description of clouds and precipitation in an ideal symmetrical cyclone probably appeared to be the best fitting pattern in the existing literature, certainly one of the most widely accepted. Data from the summer forecasting seemed to indicate that rain falls in a cyclone's warm sector. But when Jacob realized in the autumn that the rain in the warm sector most likely arose from the local effects of Norway's mountains, he concluded that cyclonic rain occurs only in connection with the lines of convergence. He simply removed that portion of the diagram covering the warm sector (Fig. 11).

Existing thermodynamic and hydrodynamic theories along with the available observations could at best only suggest, but not dictate, a unique structure for the model. Jacob was, after all, but twenty years old. Equally important, the conceptual difficulties in articulating a new cyclone model should not be underestimated. Trained primarily to think of atmospheric phenomena in terms of the statics, kinematics, dynamics, and thermodynamics of his father's project, Jacob had to *learn* to conceive three-dimensional atmospheric structures. The project did entail diagnoses and prognoses of a three-dimensional atmosphere, but in practice the latter was conceived as the sum of

---

because on 18 July, in one of the only cases in which he could clearly delineate a line of divergence accompanying a squall line, he found that the line of most rapid temperature fall coincided with the line of divergence—and not the squall line.

33. "Nogen resultater av veirvarslingen for landmænd sommeren 1918," p. 4, KUD/D, folder: "VpV Budsjett 1918–19"; copy also in JBP; a clear instance of a line of divergence leading a squall line was drawn on 18 July, giving J. Bjerknes confidence in this scheme ("Veiroversigter I").

34. J. Bjerknes, "Structure of Moving Cyclones," pp. 97–98.

35. Original drawings in posthumous papers 29a, VBP; J. Hann, *Lehrbuch der Meteorologie*, 3d ed., in collaboration with R. Süring (Leipzig, 1915).

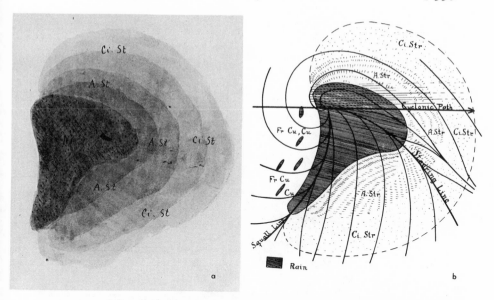

Fig. 10. Sketches of clouds associated with a cyclone: (*a*) Cloud layers depicted with water paints, (*b*) Schema of cloud layers imposed on a diagram of wind flow in a cyclone. Ci. St(r). = cirrostratus; A. St(r). = altostratus; Ni. = nimbus; Fr. Cu. = fractocumulus. From VBP, posthumous papers 29a.

several two-dimensional surfaces, or layers, in the atmosphere. Actual weather phenomena were correlated with the two-dimensional geometric structures arising from analyses of the individual variables. Clearly, the entire summer forecasting endeavor and the impressive findings it provided were extensions of the Bjerknes school's well-defined research program, the primary referent for the 1918 undertakings and the source for constructing meaning and understanding. Whether Vilhelm pressed his son to integrate the new findings into the earlier ones cannot be ascertained; he himself was first to employ language and analogies from *Kinematics* in an early report of Jacob's progress.[36] Surely he and Jacob still hoped the forecasting work would help certify the validity and practicality of the group's rational methods.

During the fall and winter 1918–19, Vilhelm Bjerknes and his assistants looked closely at this preliminary cyclone model and took

36. In a report to the Carnegie Institution, dated 18 August 1918, V. Bjerknes refers to "propagating warm waves" and "propagating cold waves" that were associated with lines of convergence in *Dynamic Meteorology and Hydrography*, 2:60; these possess the gross features of the steering and squall lines, with the exception that no surface of discontinuity is portrayed in the vertical.

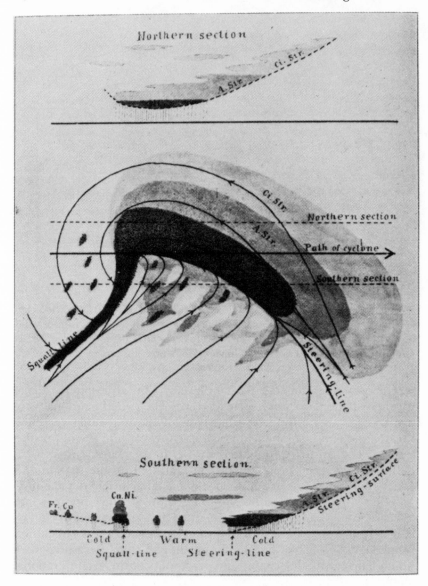

Fig. 11. Clouds and rain associated with a cyclone, revised sketch. In this sketch, which includes vertical cross-sections for points north and south of the cyclone center, Jacob Bjerknes retained the configuration of the lines of flow but removed the rain from the cyclone's warm sector (compare with Fig. 10b). From J. Bjerknes, "On the Structure of Moving Cyclones," p. 4.

stock of their institutional arrangement and their prospects for continuing with research and forecasting. In his first year back in Norway, Bjerknes had turned a number of contingencies into opportunities. While exploiting the immediate situation, he had kept in mind the long-term prospects for transforming atmospheric science. He grasped the war's profound effect on meteorology and the postwar era's promise of yet further challenges and opportunities. Like virtually all his colleagues, he anticipated regular widespread commercial flying and its enormous impact on meteorology.

PART IV

APPROPRIATING
THE WEATHER
FOR COMMERCE:
A METEOROLOGY FOR
THE POSTWAR ERA (1919)

[7]

Meteorology and the Advent of Commercial Aviation

B Y the time peace was restored, meteorologists understood that their science would be part of any future war effort. Widespread expectation in Europe that air power, long-range artill-ery, and mobilization of entire national economies would henceforth be features of war suggested that national weather services would not be reduced to prewar organizational and budgetary levels. In an ad-dress to the Royal Meteorological Society, its president, Napier Shaw, put it thus: The weather's caprices had previously been irrelevant for the military, but now,

> the whirligig of time has brought in its revenges. We are no longer allowed to regard the weather as a subject of curious inquiry that can be ignored in time of war. It has been borne in upon us that weather has its influence on the production, preservation, and transport of food; that it has a bearing upon the health of the community; that floods and droughts, sunshine and storm, such trivial circumstances as low clouds and fog, have their effect upon operations of offence and defence. . . . In these days it is realised that a gun cannot be aimed properly without more knowledge of the atmosphere than was contained in the accepted text-books of gunnery. An aeroplane cannot wisely start without consid-eration of the prospects of its finding suitable weather for landing. The engines that have to work at high levels show a very keen appreciation of meteorological science. The projectiles of the gun which bombarded Paris at long range are reported to have made their way through regions of the atmosphere explored only by the most inquisitive meteorologists.[1]

1. William Napier Shaw, "Meteorology: The Society and Its Fellows," *Q JRMS* 45 (1919), 97.

This understanding of weather as a resource for rational military operations suggested new possibilities for meteorology in peacetime. Weather, except for extreme conditions, had largely been ignored in political economy. (Climate—based on statistical patterns—had been used for planning sailing routes, agricultural purposes, and public health efforts.) Protecting sailing ships from storms had been the major reason for establishing telegraph-based weather services during the midnineteenth century. As steam replaced sail and the size and stability of ships increased, the need even for such services diminished.[2] Now the war had taught the lesson that weather was important for an efficient economy: "The great use of meteorology in warfare has shown that there are large possibilities of extending it much more thoroughly into almost every line of human endeavor."[3] It was time for meteorologists to capitalize on the wartime experiences and forge vital national disciplines, and commercial aviation posed the greatest initial opportunity and challenge.

Even before war's end, many Europeans expected aviation to contribute significantly to postwar commerce, transportation, and of course, military preparedness. Aeronautics authorities claimed that although most of the technological progress in aviation during the war had been in developing fighter planes, not aircraft for transporting materials and passengers, these advances in construction would place the latter within easy reach. Although the one- and two-seat reconnaissance and fighter planes were unsuitable for commerce, the heavy bomber, developed late in the war for tactical bombing and raids on cities, could be a model for civilian transport. Many aeronautics spokesmen asserted the airship would dominate postwar commercial aviation—or at least provide a means of long-distance, heavy-scale operations. The Zeppelin airship had been developed enormously in virtually every significant characteristic, especially in its ability to carry heavy loads. The Allies had brought down enemy airships behind their lines and learned to construct giant rigid airships based on the successful German design. Commentators anticipated giant airships in commerical use within a few years.[4]

2. W. H. Dines, "Meteorology and Aviation," *Nature* 99 (26 July 1917), 424–426.

3. "The American Meteorological Society," *MWR* 47 (1919), 875, comment on the founding of the society in 1919; see also, on the expectations for meteorology after the war, Shaw, "Society and Its Fellows," pp. 95–111; Henri Deslandres to H. H. Hildebrandsson, 26 January 1918, HHH, notes that the advances in military aviation will stimulate meteorological investigations.

4. Charles Harvey Gibbs-Smith, *Aviation: An Historical Survey from Its Orgins to the End of World War II* (London, 1970), pp. 172–79; "Technical Development, Balloons and Airships, 1914–1919;" *Aircraft Year Book 1920* (New York, 1920), p. 135; Douglas H. Robinson, *Giants in the Sky: A History of the Rigid Airship* (Seattle, 1973), pp. 250–315.

Much of this enthusiasm was, of course, ill-founded. In spite of wartime advances, many airplanes were still frail and even apt to fall apart in midair; airships, as experience later proved, were still vulnerable to a variety of unfavorable weather conditions. Economic difficulties were also grossly underestimated. At the time, though, enthusiasm abounded for a great new age of communication. Supporters of aeronautical development often said that aviation would link the nations of the world in bonds of commerce and travel the way railroads had tied together individual nations. Efforts began after the war to realize this goal.

After the armistice, civil airlines came into being and began operations. In February 1919 the German Air Transport Company (Deutsche Luftreederei) began passenger service connecting Berlin, Leipzig, and Weimar, while the German Zeppelin transport company, DELAG, resumed passenger service in August between Berlin and Friedrichshafen. In Britain the Aircraft Transport and Travel Limited and the Handley Page Transport Company began daily passenger and cargo flights between London and Paris in August and soon thereafter between London and Brussels. French transport companies began regular service between Paris and Brussels in March and later between Paris and London. These were but a few of the aviation activities in the first months of peace. An American fact-finding mission to Europe reported, "The development of aviation is progressing so rapidly at this time that it is difficult even for those in close touch with it to keep up with its progress."[5] Optimistic expectations for civil aviation were reinforced by the recognition that military aviation would be crucial in any future war. Wartime experience had shown that production and innovation in aircraft could not be obtained on short notice. A strong, efficient, economical military air force would have to rely on developments in commercial aircraft and on facilities and personnel from commercial enterprises.

In the immediate postwar period aviation's extensive meteorological needs received considerable attention, too. Despite extensive wartime experiences, meteorologists and aviators were still uncertain of needs and means. Most aviators did not understand the difficulty of predicting the weather and often expected precise forecasts of conditions several days in advance. Some pilots liked to believe their aircraft strong enough to fly in virtually all weather. Those

5. "Report of American Aviation Mission," *Aircraft Year Book 1920*, p. 112; R. E. G. Davies, *A History of the World's Airlines*, pp. 11–15, referred to in Richard P. Hallion, *Legacy of Flight: The Guggenheim Contribution to American Aviation* (Seattle and London, 1977), pp. 7–8; Robin Higham, *Britain's Imperial Air Routes, 1918 to 1939: The Story of Britain's Overseas Airlines* (London, 1960), pp. 20, 29–30.

with a financial stake in commercial aviation's reliability and success took a different view: accidents resulting in loss of life could jeopardize an airline's ability to attract passengers; economic success depended on efficient and reliable delivery of mail and merchandise. For them, meteorological considerations necessarily became primary. Some authorities regarded the meteorological difficulties confronting civil aviation's growth as even greater than the technological ones.[6]

Many, if not most, meteorologists welcomed the opportunities to bolster their science and their profession: "Meteorology will be so much poorer in its instruments of research and in the stimulus given by practical usefulness, if aviation languishes, . . . it behooves meteorologists to strengthen the intimate connection between aviation and meteorology which grew out of the war."[7] Meteorology could flourish institutionally because of the extra money that would be directed to forecasting and to research on forecasting; new university professorships and departments could be made possible.[8] Of course the chance to profit from aviation's needs often was limited by the stipulations of those with political authority who were concerned with aviation.

In Britain, for example, aviation's needs led to institutional change, at times even against the meteorologists' own desires. In 1919 the Air Ministry, through its Department of Civil Aviation, assumed control of the Meteorological Office. From its creation in 1867, through 1919, the "Met Office" had been independent of ministerial control on account of its various scientific duties and had been responsible only to Parliament through the Treasury. Shaw had resisted the Air Ministry's attempts, beginning in 1918, to place the weather service under its control, but to no avail. The Met Office's autonomy was whittled away. By March 1919 the War Cabinet's Research Committee recommended that the Met Office move to a location in nearest phys-

6. Lt. Francis W. Reichelderfer, "The Present Meteorological Needs of Aeronautics," *U.S. Air Services* 10, n. 8 (1922), 22–28; R. Bernard, "The Weather and Commercial Aviation," *Aeronautical Engineering* (supple. to *Aeroplane*), 26 November 1919, p. 1906; discussion of talk by Lt.-Col. H. G. Lyons, "The Supply of Meteorological Information for Aeronautical Purposes," *Aeronautical Engineering* (supple. to *Aeroplane*), 17 September 1919, pp. 1094–96.

7. Ernst Gold, "Meteorology and Aviation," *Journal of the Scottish Meteorological Society*, 18 (1919), 68–76; quotation on p. 68.

8. Napier Shaw to V. Bjerknes, 3 September 1920, VBP; Alfred Angot to Hildebrandsson, 22 June 1919, 27 January 1921, HHH; V. Bjerknes to H. U. Sverdrup, 15 October 1921, copy, VBP; "Comité d'action des services météo, français 1917–1918" and "Réorganisation des services météo, 1917–1921," Bureau central météorologique, Archives nationales, Paris, F1713594. Aeronautics, for example, prompted the creation of a professorship in meteorology at Imperial College, London, and an institute for terrestrial physics with emphasis on meteorology in Paris.

ical proximity to the Admiralty, the War Office, and the Air Ministry, and that their representatives should sit on the Met Office's board of management. In October the undersecretary of state for air restructured the Meteorological Committee, which oversaw the Met Office's activities, placing the controller general of civil aviation as president and giving the Air Ministry two additional representatives. Henceforth, the Met Office was to receive its funds from the Air Ministry. In November the Forecasting Division was moved to the Air Ministry, where instead of attempting to satisfy the needs of various interests, it had "for its primary and immediate object the satisfaction of the requirements of aircraft."[9]

Similarly, the French military, which included aviation in its administration, continued its control over the national weather services in peacetime. The aerological observatory at Trappes remained under military control; even appointments to academic positions in meteorology apparently did not escape the military's influence. This general "militarizing of meteorology," to use Vilhelm Bjerknes's expression, accompanied to various degrees the expansion of weather services and of meteorological research in many countries.[10] Some, meteorologists, for example, Lewis F. Richardson, protested and even abandoned the profession because of this interconnection with the military.[11] Many meteorologists, however, viewed the development of predictive methods and the study of problems relevant to flight as a challenge both unavoidable and promising in its scientific and profes-

9. *Fourteenth Annual Report of the Meteorological Committee to the Lord Commissioner of His Majesty's Treasury: For the Year Ended 31st March, 1919* (London, 1919), pp. 15–19; *Fifteenth Annual Report . . . : For the Year Ended 31st March, 1920* (London, 1920), p. 86–88; quotation from William Napier Shaw, Forecasting Weather, 3d ed. (London, 1940), p. 419.

10. V. Bjerknes to Sverdrup, 15 October 1921, copy, VBP; Bjerknes noted that one of the major transformations in atmospheric science since the war had been the "militarizing of meteorology" as a consequence of the discovery by the military of meteorology's usefulness and necessity for commercial and military aviation. According to Bjerknes, this takeover of national weather services by military-aviation interests resulted in directing unprecedented financial resources to these institutions, but the meteorologists' freedom to use the money and to operate as they desired had been greatly curtailed. See also H. Solberg to V. Bjerknes, 14, 20 May 1921, VBP; Angot to Hildebrandsson, 27 January 1921, HHH. Bjerknes, like most scientists of his generation, seemingly maintained a deep-rooted faith in scientific progress as a means for improving the cultural and material standards of society. Also like most scientists, he did not directly confront the moral issues raised by the war.

11. On Richardson, a Quaker and visionary pioneer in numerical weather forecasting, who could not accept working for the Met Office—even on theoretical investigations—once the Air Ministry had assumed responsibility for its operations, see Oliver M. Ashford, *Prophet or Professor? The Life and Work of Lewis Fry Richardson* (Bristol and Boston, 1985).

sional rewards, at least when institutional and organizational structures might allow sufficient opportunity for inquiry.

Weather and Forecasting Redefined

The tension that had developed during the war between the state of meteorology and the tasks confronting it increased once forecasters and researchers confronted the projected needs of commercial aviation. Military weather services had revealed their inadequacies for assisting aerial operations. Storm warnings for coastal regions were crude with respect to location and duration. A ship, after all, could remain at sea if her port was closed by dense fog; she could often be safely maneuvered out of danger of an oncoming storm if a warning signal on the coast could be picked up in advance. Flying during the war made different demands. Forecasters had to respond to inquiries such as Will the sky clear before midnight? and Will there be fog at such and such a place some fifty-to-sixty miles away at 2:00 AM?[12] Again and again it was brought home to meteorologists that the basic notions of a forecast had been transformed:

> In many directions the importance of the atmosphere in its many moods has been forced upon us recently, and during the past five years [military] operations in many parts of the world, and especially the rapid development of aviation, have made new demands upon all those who make a study of weather conditions, and who devote themselves to the investigation of the numerous and varied problems which the atmosphere provides. More detailed information is asked for, greater precision is required, and the regions over which such information is wanted are ever growing wider.[13]

In addition to forcing meteorologists to change the temporal and geographic precision of their forecasts, aviation also began forcing them to see—to define and classify—weather in much greater detail. Weather, as reported by observers in the field and as specified in predictions, had been imprecisely divided into categories such as fair, cloudy, rain or snow, and thunder. The internationally accepted code for weather observations included a single-digit (0–9) for describing weather. Restricting the classification to but ten varieties of weather,

12. Col. H. G. Lyons, "Meteorology and Aviation," *Aeronautics* 16 (5 February 1919), 158, based on a paper read at the Royal Society of Arts on 22 January, "Meteorology during and after the War"; Colonel Lyons was then acting director of the Meteorological Office. See also Dines, "Meteorology and Aviation."

13. Lyons, "Supply of Meteorological Information," reprinted in *Aeronautical Journal* 23 (1919), 397.

of which three were degrees of cloudiness, arose in part from the technology and economics of weather forecasting. Using telegraphy to transmit weather data meant that every piece of information added to both complexity and cost of a telegram. Moreover, no economic or scientific reason had as yet compelled forcasters to distinguish weather into finer categories.

Flying during the war and immediately thereafter showed the internationally agreed upon classifications to be wholly inadequate. In his article "Aerial Navigation and Meteorology," British meteorologist Ernst Gold underscored the problems arising from using a code that does not distinguish between light and heavy precipitation or types of clouds:

> The need for an extension of the existing one-figure code [for observations of "weather"] has been apparent for a long time. A meteorologist at headquarters requires from a reporting station sufficient information to enable him to say with precision and certainty what the weather was at the station at the time of report. With the pre-war code for international exchange this was not done. A few drops of rain or a little drizzle were reported by the same figure as the most torrential downpour. A few flakes of snow or some fine ice-crystals were reported by the same figure as the heaviest snowstorm.... A sky nearly covered with thin, white clouds at 20,000 ft. or 30,000 ft. was described by the same figure as the darkest, gloomiest day of the year.[14]

Commercial and military aerial operations now required seeing, reporting, and predicting differentiations. In no uncertain terms, Shaw, head of the international meteorological organization, pointed out, "The old international telegraphic [code] . . . does not satisfy the requirements of aerial navigation."[15] Gold and others proposed a new code that distinguished one hundred types of weather, the "ww" code.

Regular flying also forced meteorologists to include in forecasting work previously ignored phenomena such as cloud forms (in contrast to merely the amount of sky covered by clouds) and visibility. Once airplanes and airships became commercially and militarily important, their great speeds and vulnerability when landing endowed value to the atmosphere's transparency. Efforts to define scales for observing and predicting changes in visibility and in low-, medium-, and high-level cloud forms became common after the war:

14. Ernst Gold, "Aerial Navigation and Meteorology," *Nature* 105 (19 August 1920), 776.

15. Shaw to "My dear Colleagues" [heads of national weather services and leading meteorologists], 3 June 1919, HHH and NMI.

A few short years ago ceiling [the height of the lowest layer of clouds or other obscuring phenomena] and visibility were scarcely given a thought in meteorology. Now they are among the most important working tools of the meteorologist whose duty is to tell the pilot what the weather is and is going to be.[16]

It follows that visibility is one of the most important meteorological conditions to the aviator. . . . At the present time visibility, except in the cases of dense fog, is a thing the general public is little interested in and unfortunately visibility is one of the difficult things to forecast.[17]

Aviators needed to know these detailed weather conditions along their flight routes; conditions could change markedly during the course of a day. Making and collecting observations once or twice daily, was not sufficient. Meteorologists readily agreed on the need for at least four observation periods per day; extra observations could be coordinated with specific flights. Meteorologists who may have wanted such frequent collection and analysis of data in the past had, of course, lacked the social or economic justification to convince national governments to foot the bill.

Finally, meteorologists had also deliberated upon the actual predictions themselves during the war and come up with a new type of forecast, called a short-period prediction: detailed weather conditions for a specific location for the next few hours. Before World War I neither meteorology's social tasks nor the available communication technologies had made this type of forecasting a realistic goal for national weather services. For aviation, such forecasts became absolutely necessary. In these new short-period forecasts meteorologists aimed to specify cloud cover, weather, visibility, surface wind, and wind conditions at 1000 and 2000 meters above the surface, at a specific location, for the next two-to-four hours. Traditional forecasts for the coming twelve to twenty-four hours would now contain similarly enriched detail, but less specific than that in the short-term predictions.

The expected increase in quantity and frequency of exchanges of weather information necessitated changes in communication systems; for both sending observations to central forecasting bureaus and disseminating predictions. Short-term forecasts had to be transmitted directly to air fields as quickly as possible, by telephone and, increasingly, by wireless telegraphy: "The Royal Air Force required

16. Willis Ray Gregg, *Aeronautical Meteorology*, 2d ed. (New York, 1930), p. 169.

17. Reichelderfer, "Present Meteorological Needs of Aeronautics," p. 24; see also F. J. W. Whipple, "The Horizontal Range of Vision as a Meteorological Observation," *QJRMS* 48 (1922), 85, 88.

forecasts of weather for short periods, which they could use for their reconnoitering and bombing squadrons . . . for all this the most rapid means of transmission is essential."[18] "The most important thing from the aviation point of view was that the information had to be got through very quickly."[19] International exchanges of weather information grew in significance once commercial flying began. Both explicitly and implicitly the drawing-up of the new forecasting goals and the increased exchanges of information depended upon rapid communications.

Although in 1919 meteorologists and aviators could agree that forecasting had to change, discussion focused primarily on mechanical changes in routine and classification; the theoretical or conceptual bases for change received much less attention. Could the prewar forecasting methods based on formalistic empirical rules related to delineating areas of low and high pressure be used to arrive at precise short-term predictions? Would the large increase in the amount and type of meteorological data, and the need to analyze these data to provide forecasts for aviation, result in a new understanding of the atmosphere? Shaw and some other meteorologists, claimed that wartime experiences revealed the limitations of forecasting based on empirical rules; meteorology's vastly increased social importance required that "we must depend upon our knowledge of the dynamics and physics of the atmosphere."[20] But how? Others sought expansions of their formal rules on the basis of the enriched observations, such as the French "*système nuageux*" based on cloud forms.

Naturally, meteorologists' temperaments and education and their institutions' organizational and operational traditions figured in individual attitudes. In general, institutional conservatism continued even after the war; modifications and ad hoc additions to existing methods prevailed at first. Meteorologists understood that a new epoch had begun; most of them, however, required time to appreciate just how comprehensively they needed to change their predictive methods, atmospheric models, and theoretical outlook if they were to begin resolving the tensions that existed between the state of meteorology and its social purpose.

18. Lyons, "Meteorology and Aviation," p. 158.
19. Maj.-Gen. Brooke-Popham comments on Lyons's "Supply of Meteorological Information," *Aeronautical Engineer*, p. 1094.
20. Shaw, "Meteorology," p. 102.

[8]

The Start of the Bergen School: Institutional, Methodological, and Conceptual Change

DURING the fall and winter 1918–19, Vilhelm Bjerknes reflected on the significance of the summer's experiences. Expressing great pleasure, he claimed that this effort, the first in which he and his assistants had made "real contact with practice [*praxis*]," had a very successful outcome.[1] "Professional authoritative sources" had lectured him that the "course" of his research was impractical and never would amount to anything, but suddenly it appeared that his son had "picked up" during the daily forecasting the cyclone's "most important secret."[2] He was now convinced that close interaction between dynamic meteorology and daily practical forecasting was essential for the progress of both. One could inform the other. As he saw it, practical methods could be established based on a physical understanding of the atmosphere; concepts and insight arising from analyzing weather maps could provide direction and assistance in creating a physics of the atmosphere.[3] Bjerknes's commitment to interaction also reflected his understanding that a permanent forecasting service connected to his institute would be crucial to his professional aspirations at home and abroad. In 1917 and 1918 he and his assistants had experimented with institutional arrangements and forecasting practices; in 1919, Bjerknes attempted to lay down a stable foundation—institutional, conceptual, and methodological—on which he could develop a school of meteorology for the postwar era.

1. V. Bjerknes to Svante Arrhenius, 27 December 1918, SAP.
2. Ibid., 3 November 1918.
3. V. Bjerknes to KUD, 1 March 1919, KUD/D, folder: "Værvarslinga på Vestlandet," from file: "Værvarslingstjeneste (eldre dokumenter) 1918–23."

Bjerknes by now recognized that the Bergen Museum's Geophysics Institute was unlikely to foster a vital research school. The original plan for Bjerknes to join an already active geophysics milieu consisting of the Oceanographic Institute and the Bergen Observatory was dead: these had become inactive during the war. Progress toward establishing a university in Bergen was disappointing; little of the money pledged was forthcoming. Soaring inflation cut sharply into the value of the money that had been donated. Bjerknes was disillusioned: "Had I suspected beforehand that this was the situation into which I would get involved, then I would never have thought of accepting the call to Bergen."[4] He developed a bitterness toward the leaders of the museum and many of the leaders of Bergen's establishment. As Bjerknes saw it, he was but a prize catch to be put on display as an example of Bergen's cultural pretensions.[5] He had expected that a privately funded institution, not unlike the Stockholm Högskola, was to be erected. The plan actually entailed using private funds to begin such an institution but then turning to the state to achieve and maintain the endeavor. Bjerknes did not believe this strategy was beneficial for the nation's scientific and cultural development because the university in Christiania and the technical college in Trondhjem were already having difficulty obtaining funds and qualified personnel. What is more, Bjerknes did not think the plan would succeed.

One possibility for shoring up the Geophysics Institute entailed a plan for creating in Norway institutions for assisting in the reestablishment of cultural cooperation among nations. Believing that the neutral nations could promote the resumption of international exchange of ideas, Fredrick Stang, a prominent jurist and member of the Storting, proposed founding an institute for the comparative study of cultures in Christiania and expanding the Bergen Geophysics Institute into an international center for research and study. Scandinavian geophysicists in general acknowledged the strong international orientation of their fields and the desirability of taking initiatives to resume European cooperation; this might be their opportunity to assume a more central role in international science.[6] Stang's plan fit Bjerknes's goals perfectly, but Bjerknes doubted the

4. V. Bjerknes to Fridtjof Nansen, 9 January 1919, FNP.

5. Ibid.; V. Bjerknes to Arrhenius, 5 January 1924, 18 February 1925, SAP; V. Bjerknes to Sem Sæland, 21 November 1923, 10 February 1925, SSP. Of course in Bjerknes's public speeches none of this disappointment is evident.

6. V. Bjerknes to Nansen, 9 January 1919, FNP; Nansen to V. Bjerknes, 6 February 1919, VBP; V. Bjerknes to P. B. Vogt (Norwegian Legation in London), 17 February 1919, copy, VBP; closing comments by Helland-Hansen and Bjerknes in *Skandinaviska geofysikermötet i Göteborg den 28–31 augusti 1918*, ed. O. Nordenskjöld and H. Pettersson (Göteborg, 1919), p. 63.

Bergen half of the plan would be realized, for he was skeptical about the prospects for fund raising. Museum president Johan Lothe, a major force behind the plans for a Bergen university, believed that the Stang plan might divert attention and funds from the university; Bjerknes suspected—rightly—that Lothe and others would sacrifice the Stang plan.

Bjerknes saw therefore only one certain way to achieve a stable foundation for his work and for the Geophysics Institute: only by accepting "serious tasks" could the institute survive in the future; weather forecasting could provide "serious practical and theoretical research work in connection with it."[7] He believed practical and theoretical research in weather forecasting was also most likely to help the institute achieve an international reputation. Bjerknes admitted that involvement in practical forecasting, which had so far proved successful, might mitigate his regret at coming to Bergen: "Through my dealings with weather forecasting, I have now secured a goal toward which to steer."[8] Bjerknes felt that Bergen was indeed the right place geographically to obtain a deeper understanding of atmospheric and other related geophysical processes and that the time was also right for "theory and practice" to interact: "This is a situation which I personally must attempt to exploit to the fullest."[9] And there was the possibility of state support.

Much of the rationale he used in 1918 for petitioning for funds was still valid in 1919. Even after the end of the American export boycott and the war, food prices and farm labor costs remained extremely high. Government efforts to support Norwegian agriculture continued to grow.[10] Regulations and subsidies to promote "internal colonization" of new land involved the state in expensive engagements for increasing the national production of food stuffs. Bjerknes could again claim that even a one percent increase in crop yield, or a one percent savings in labor costs, achieved through accurate weather forecasting, would pay for the entire nation's weather services for a year.[11] He saw, too, that the general wave of enthusiasm for national

7. V. Bjerknes to Nansen, 9 January 1919, FNP.
8. Ibid.
9. Ibid.
10. Hans Trøgstad, "Jordbruksproduktionens økning i krigsaarene," *Norsk Landmandsblad* (1919), 296–97; *NOS XII 245*, Table 100, "Working Results of Farms" and Table 276, "Index of Agricultural Prices, 1909–14 to 1956–57."
11. V. Bjerknes, "Betænkning om den norske veirvarsling og dens utvikling," 1 September 1919, KUD/D, folder: "Værvarslinga på Vestlandet," file: "Budsjettsaker 1916–46" (hereafter cited as "VpV"/"Budtsjettsaker"). V. Bjerknes, "Om vær- og stormvarslinger og veien til at forbedre dem," *Teknisk Ukeblad* 38 (1920), 300.

self-sufficiency and for science as a means was still gaining in momentum. Prime Minister Knudsen's liberal party had even included in its 1918 electoral platform the development of research capabilities to help place Norwegian commerce on a scientific basis.[12] Science as a means to help rationalize state-supported production seemed to have growth potential in an otherwise traditionally bleak national research landscape.

Another reason Bjerknes could seek a permanent weather service for West Norway was his assumption that additional manpower and money would have to be invested in forecasting for commercial aviation.[13] Enthusiasm for flight in Norway was enormous. In correspondence among themselves and in reports to the Ministry of Church and Education, Bjerknes and his meteorological colleagues in Norway repeatedly alluded to aviation's expected needs for improved weather services to justify increasing the budget for forecasting and forecasting-related activities. From the vantage point of the immediate postwar optimism about aviation's worldwide future, Bjerknes could well believe that the expansion of resources for weather forecasting was certain.

Bjerknes campaigned to make the forecasting service a permanent year-round institution and to bring it under the jurisdiction of his institute. The Bergen Meteorological Observatory could be expected to claim responsibility for daily forecasting during the coming summer and for any eventual year-round forecasting service. Not only might the physical distance between the two institutions pose a hindrance to intimate contact between "theory and practice"; Bjerknes and his assistants actually had no formal authority for decisions and operations at the observatory.[14] B. J. Birkeland, its leader, had allowed the Bjerkneses a free hand to run the summer 1918 service; this hospitality could not be guaranteed for the future, especially as Bjerknes and Birkeland did not get along at all. Bjerknes decided to claim the forecasting for his own institution.

In 1918 the wealthy Bergen merchant Johan A. Mowinckel had

12. John Peter Collett, "Videnskap og politikk: Samarbeide og konflikt om forskning for industriformål, 1917–1930" (Diss., *hovedoppgave*, Dept. of History, Oslo Univ., 1983), 89–90.

13. V. Bjerknes to Wilhelm Keilhau, 1 May 1918, KUD/D, "VpV"/"Værtj"; V. Bjerknes to Olaf Devik, 17 April 1919, copy, VBP; Devik to Telegrafstyret, Tromsø, 8 February 1919, copy, NMI; V. Bjerknes, "Betænkning om den norske verivarsling"; Theodore Hesselberg to KUD, 18 June 1919 and Hesselberg to Centralkomitéen for videnskabelig samarbeide til fremme av næringslivet, 28 May 1919, letter book 25, NMI.

14. V. Bjerknes to KUD, 1 March 1919, KUD/D, "VpV"/"Værtj."

Plate 2. West Norway Weather Bureau (1919). Located in the attic of the Bjerknes residence, the forecasting service provided opportunities for frequent, informal scientific discussions outside office hours. Seated on the left (from foreground to background) are Tor Bergeron Carl-Gustaf Rossby, and Svein Rosseland. Jacob Bjerknes is standing. On the right are clerical assistants Johan Larsen, Sverre Gåsland, and Gunvor Førstad. The photograph was taken during fall 1919. Copy courtesy of the West Norway Weather Bureau.

donated a large house to the museum for locating the meteorology division of the Geophysical Institute.[15] Bjerknes proposed at first that forecasting be transferred from the observatory to his institution during the spring and summer, at which time he planned to hold a course

15. Bergens Museum, *Aarsberetning for 1918–1919* (Bergen, 1919), p. 6; J. A. Mowinckel's son, ship owner and future prime minister Johan Ludwig, also gave generous donations to the institute and the weather service. The Mowinckel family was active on the Bergen Museum's governing committees. Bergen shipping magnates in general supported the forecasting service and the Geophysical Institute; *Bergens Museum 1925: Historisk fremstilling* (Bergen, 1925), pp. 466–68. Although shipping benefited from the forecasting service, its special needs did not prompt changes in forecasting practice; the forecasting requirements for aviation largely encompassed those for shipping.

in meteorology for Scandinavian students. He then proposed that the summer forecasting service be connected permanently to the Geophysical Institute and extended year-round.[16] He claimed that only by keeping forecasting in closest contact with theory could there be improvement in methods. His division of the Geophysical Institute should encompass dynamic meteorology and related practical forecasting; the Bergen Observatory should then be responsible for climatology and for making aerological observations.

Birkeland opposed these efforts.[17] He noted that the Geophysical Commission, which advised the Ministry of Church and Education to make these changes, was composed of persons partial to Bjerknes and his work. He wondered why the advice of older meteorologists at the Meteorological Institute in Christiania had not been sought. Birkeland lost; he eventually moved to the climatological division of the Meteorological Institute. Bjerknes became in July 1919 the head of a provisional new entity, West Norwegian Weather Bureau (Den vestlandske veirvasling), which in turn became in July 1920 the West Norway Weather Bureau (Veirvarslingen paa Vestlandet),"[18] with Jacob Bjerknes as chief. Although the weather forecasting service itself was formally a division of the Norwegian Meteorological Institute in Christiania, especially in matters of administration and budget, to a large extent routines and decision making were to be left to the forecasting service in its new location in the attic of the house donated to Bjerknes. There the weather service would be in direct contact with Bjerknes's institute and—during the critical first years—largely under his supervision.

New Techniques: Integrating Theory with Practice

Bjerknes undertook to meet the forecasting service's mandate to provide economically beneficial predictions. For agriculture, he called for continuing to issue individual detailed forecasts to as many specific forecasting districts as possible. Precision and detail must inform the forecasting endeavor: "This . . . is the necessary condition so that the entire [forecasting] system should be truly useful for agriculture."[19] These criteria for benefiting agriculture could be subsumed in the more comprehensive and demanding needs of forecasting for

16. V. Bjerknes to KUD, 1 March 1919, KUD/D, "VpV"/"Værtj."
17. B. J. Birkeland to KUD, 6 June 1919 and n.d. [ca. 29 December 1919], KUD/D, "VpV"/"Værtj."
18. I am using Bjerknes's spelling from the time; variations in titles arise from changes in orthography and from regional variations. *Veir, Ver,* and *Vær* all mean "weather."
19. V. Bjerknes, "Betænkning om den norske veirvarsling."

aviation. Economic advantage would accrue to both sectors if they integrated weather into operations to the greatest extent possible, if they could treat weather as a detailed and mapped-out resource. "For me it still remains as meteorology's task precisely to get away from this coarse-grained kind of forecast [that characterized meteorology prior to the war], in order to arrive at forecasts with ever sharper determinations of time and with ever subtler distinguishing [of types of weather] for the individual sections of the forecasting districts."[20]

Toward this goal, Bjerknes and Hesselberg had for the summer 1918 experiment divided the southern half of Norway into nineteen forecasting districts. For the summer 1919 forecasting effort, they proposed a yet finer division into fifty-four districts. The network of observation stations would remain at the same level of tightness, but without the navy observations stations. Replacing these would be international telegrams, some already arriving by wireless. Nine Norwegian ships that regularly crossed the North Sea had agreed to telegraph weather observations. Observations would be collected three times daily. Counting on this expanded vision to the west, the Bergen group agreed that starting in the summer 1919, they would attempt forecasting for the next day, a service farmers in West Norway and Trøndelag had requested. Next-day forecasts would have a better chance of reaching the farmers in time to help them plan their work. From both Christiania and Bergen, forecasts would be issued by 1:00 PM and be valid until 8:00 PM the next day. As in the 1918 effort, predictions would be spread by telegraph to telephone stations from which farmers could receive information.[21] At the same time the Bjerknes group considered these mechanical measures for improving the summer forecasting, it also assessed the conceptual and epistemological bases for the predictions.

Bjerknes did not want to abandon his goal of creating a physics of the atmosphere. Neither did he want to repudiate his pre-Bergen research program aiming at rational precalculations of atmospheric states. Although he tended to claim the summer's events a victory for his research program, he surely understood that his earlier assumptions about the practicality of the rational methods had not been vindicated. The impracticality of precalculating changes in the atmo-

20. V. Bjerknes to Hesselberg, 15 January 1919, copy, JBP; see the reports sent to KUD, 10 December 1918, by V. Bjerknes, J. Bjerknes and H. Solberg, and T. Hesselberg, respectively: "Forslag til forbedring av veirvarslingen for landbruket sommeren 1919"; "Nogen resultater av veirvarslingen for landmænd sommeren 1918"; "Beretning om den utvidede veirtjeneste for jordbruket paa østlandet sommeren 1918," all found in KUD/D, folder: "Veirvarslinga på Vestlandet," file: "Budsjetter 1918/19–1923/24."

21. *NMIÅ 1918–1919*, p. 13.

sphere was clear from the inaccessibility of the technological basis for using the pre-Bergen methods: direct aerological observation of the winds aloft was virtually precluded by the extremely high postwar prices of the rubber sounding balloons. Nor was technology yet available for providing a vertical profile of several meteorological variables: balloon-carried recording instruments that return to earth by parachute could not be retrieved and returned quickly enough for use in forecasting. Bjerknes and his assistants devised a system of forecasting that permitted them to compensate for the lack of direct aerological observations while they strove to meet the goals of using physical theory as well as being economically useful.

As a guiding principle for the forecasting, Bjerknes proposed a new approach, based on his original objective of rationally predicting changes of atmospheric states, only now the application of physics would be made *qualitatively*.[22] Ideally, he claimed, the prognostic problem could be solved mathematically with the help of mechanics and thermodynamics; practical meteorological methods could perhaps be devised that qualitatively use the knowledge implicit in the mechanical and thermodynamic equations.

Bjerknes had originally called for defining atmospheric states by seven meteorological variables; changes in these variables would be calculated with appropriate equations. "The weather changes because the masses of air *move,* and *change their internal state.* All is done if we are able to determine their new positions and the internal state in which they arrive," he wrote in 1904. A qualitative mathematical approach required that individual air masses and their changes of state therefore be defined in terms of the interplay among all the meteorological elements. Noting the mathematical requirements for determining the system of hydrodynamic and thermodynamic equations entailed in atmospheric change, he claimed that to ignore any one element would leave the problem mathematically indeterminate.[23] No doubt this program declaration was at least partly a rhetorical means for Bjerknes to claim continuity with the pre-Bergen endeavors and legitimize forecasting as exact scientific practice. Still, when translated into actual forecasting techniques, the goal of qualitative applications of physical theory meant that rather than analyze just the field of pressure or of wind, all the basic meteorological variables would have to be analyzed in relation to one another. Each observa-

22. V. Bjerknes, "The Structure of the Atmosphere When Rain is Falling," paper delivered in London, 7 November 1919, and pub. in *Q JRMS* 46 (1920), 119; see also idem, "Om forutsigelse av regn. Foredrag ved Videnskapsselskapets aarsmøte i Kristiania den 3. mai 1919," *Naturen* 43 (1919), 322–23.

23. V. Bjerknes quotes from his 1904 article in "Om forutsigelse av regn," p. 323.

tion would have to be accounted for by a physical explanation consistent with the other observed weather phenomena. Of course, analytic techniques to realize this goal would have to be developed, a task not unlike the construction of a new instrument.

To arrive at methods for analyzing weather observations that could satisfy the scientific and economic aims of the forecasting service, Bjerknes and his assistants altered the manner in which they conceptualized atmospheric phenomena, which in turn entailed changes in the manner in which they interpreted data. They replaced the two-dimensional geometric models of atmospheric phenomena, based on kinematics of the wind flow, with three-dimensional models of physical weather-carrying systems in the atmosphere. They accomplished this change by introducing a method of interpreting weather maps in conjunction with direct observation of the sky.

Conceptual Change

The Bergen meteorologists understood that to establish a forecasting service that could be significant for agriculture and aviation, weather phenomena must be localized in time and space to the greatest degree possible. Both for aviation and for the ideal of a mathematically defined three-dimensional atmosphere, an analysis of surface data alone was not acceptable. Jacob Bjerknes's new cyclone model suggested a strategy for developing a conceptual basis with which they could establish a systematic forecasting practice. Important weather phenomena could be associated with a cyclone's discontinuities. In practice, however, the lines of convergence proved troublesome; to use them in forecasting, exceedingly precise measurements of wind from a very dense network of observation stations would be necessary. To base a forecasting system on locating lines of convergence was, at best, dubious. Moreover, considering weather phenomena in three dimensions would be difficult while focusing on these constructs that exist as two-dimensional singularities in the horizontal wind flow. So, rather than focus on the cyclone's lines of convergence as the object of analysis and develop atmospheric models on the basis of geometric kinematic ones, Vilhelm Bjerknes and his assistants began in 1919 to focus instead on the actual weather phenomena that had been associated with each line and to comprehend them with models of the supposed three-dimensional "physical" atmospheric structures responsible for the observed weather.[24] In an August 1918 lecture, Bjerknes related

24. V. Bjerknes, "The Structure of the Atmosphere," pp. 120–28. The term *physical* was used to emphasize the actual weather phenomena in contrast to their geometric patterns in the field of motion.

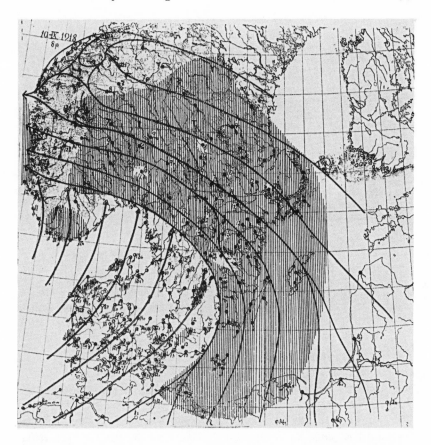

Fig. 12. Broad rain-stripe (1919). The gray area denotes rainfall. From V. Bjerknes, "The Structure of the Atmosphere When Rain Is Falling," *QJRMS* 46 (1920), 123. (Fig. 12–15 were first presented in Bjerknes's May 1919 lecture, "Om forutsigelse av regn," *Naturen* 43 [1919]).

how numerous lines of convergence had swept across the coast during the summer. In 1919 he described these events as follows: "A great number of rain-stripes . . . swept over our country."[25]

The shift toward working with weather phenomena rather than abstracted atmospheric variables can be noted in Bjerknes's description of the rain associated with steering lines. Starting with a "broad rain-stripe" as seen on the map (Fig. 12), he noted that the warmer air appeared to be blowing toward and perpendicular to the rain band itself and to the cold air in the rain band. There is no discussion of

25. Ibid., p. 126.

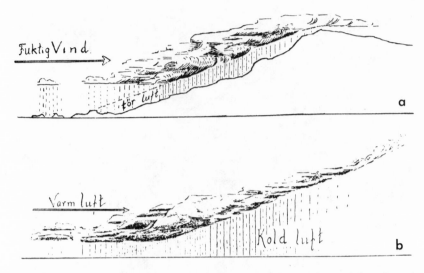

Fig. 13. (*a*) Wind blowing against a mountain; (*b*) steering surface (1919). *Fuktig* = moist; *tör* = dry; *luft* = air. From V. Bjerknes, "The Structure of the Atmosphere When Rain Is Falling," pp. 121, 123.

lines of convergence or divergence, nor of lines of flow. To make it possible to comprehend three-dimensionally what they observed on their surface-level map, he used an analogy. He likened the situation to one in which moist air blowing in from the sea toward the mountain ranges in western Norway is forced to rise up over the mountains, thereby producing a characteristic pattern of clouds and rain (Fig. 13). But, wrote Bjerknes, the mountain of cold air does not have the same withstanding power as the Norwegian mountains and therefore moves under the pressure exerted by the advancing warm air.[26]

In a similar fashion the narrow rain-stripe that had been associated with the squall line was now explained physically by cold dense air ploughing under and lifting violently the lighter warm air lying in its path, thereby forming a narrow band of showers and thunderstorms (Fig. 14)[27] Although he still used Schmidt's laboratory model of a squall line as a guide in constructing a vertical profile, Bjerknes ceased to structure the model around joined lines of divergence and convergence as in his and Jacob's 1918 articles. The new depiction of a cyclone reflects the change in the mode of developing models of the atmosphere.

26. Ibid., pp. 122–23. These thoughts are derivative from Jacob's 1918 work, only now they are given greater explanatory prominence.
27. Ibid., pp. 125–26.

Fig. 14. Squall surface (1919). The peculiar shape of the "head" of the advancing cold air is based on Schmidt's laboratory model of line squalls. Only after the Bergen group obtained European aerological observations made during the war could the squall surface's upper extent and structure begin to be determined. From V. Bjerknes, "The Structure of the Atmosphere When Rain Is Falling," p. 125.

In the new cyclone model (Fig. 15) the lines where cold and warm air meet are no longer lines of convergence. Although the expressions steering line and squall line are still used, their meaning and significance have been altered. Lines of convergence could no longer exist and function in the Bergen meteorological system as they had previously. If the endeavor was to be defined by explaining and predicting the movement and the change of state of three-dimensional bodies of air, then physical boundaries between two such air masses must, in principle, exist. In the earlier studies, lines of convergence consisted of discontinuities in a two-dimensional wind field, and although these structures could be said to represent discontinuities in the wind between differing air masses, they certainly could no longer hold any privileged status, for discontinuities could be found in other variables as well. By envisioning three-dimensional masses of air, the surface boundary could be defined as an entity in itself: the three-dimensional physical boundary between air masses across which occur discontinuities in several variables.

Analyzing and conceptualizing in two dimensions had impeded the articulation and integration of three-dimensional concepts into the explanatory scheme. In the new depiction of their cyclone model, the wind no longer flowed into and merged with the lines of convergence; rather, on a horizontal projection, the wind was depicted as blowing almost perpendicular to distinct boundary lines. These lines seem to have been endowed with a reality separate from the wind flow; they represent the intersection of the three-dimensional boundaries with the ground. What Jacob had called steering and squall surfaces (later, warm and cold fronts) could signify the physical boundaries between air masses. These structures replaced two-dimensional lines of convergence as the primary focus of practice and inquiry. In August 1918

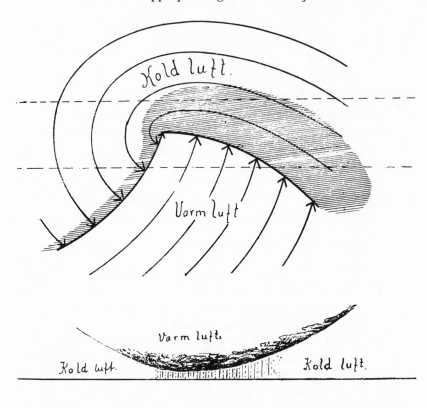

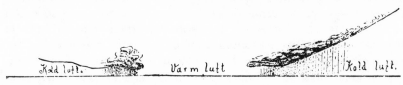

Fig. 15. Bergen cyclone model (1919): (top) horizontal projection; (middle) vertical projection north of center; (bottom) vertical projection south of center. From V. Bjerknes, "The Structure of the Atmosphere When Rain Is Falling," p. 128.

after the initial breakthroughs Bjerknes had written, "The investigation of these two characteristic lines of convergence will evidently be important not only for the theory of cyclones, but even for practical weather forecasting."[28] One year later he wrote, "The atmosphere is crossed and recrossed by surfaces of discontinuity, separating from each other masses of air having more or less different velocity and

28. *CIWY, 1917–1918*, p. 301.

different physical properties. . . . Almost every change of weather is due to the passage of a surface of this kind."[29]

For these discontinuities to be immediately useful in the forecasting endeavor and to convince other meteorologists of their reality and significance, the Bergen meteorologists knew they had to specify observational requirements and analytic techniques with which to reproduce discontinuities in practice: "For promoting weather-forecasting it will be of high importance to arrange the observations so that the formations of the discontinuities can be detected at an early state, and their propagation followed as accurately as possible."[30] Of course they would eventually also want to reproduce in theory these structures, their precise nature, and how they are formed and maintained; but that task could wait until they had tackled the more pressing challenge of stabilizing the models within a system of weather forecasting.

While defining the surfaces of discontinuity, the Bergen group considered criteria for appropriate forecasting practices. A dense network of observations in which various weather variables were measured on a sufficiently fine scale could aid in identifying the surfaces of discontinuity where they intersect the ground. Similarly, delineating the geographic extent of rain patterns in a dense network could also help show the location of the discontinuities, at least near the ground. To incorporate these tactics they introduced into the instructions and codes for weather observers a number of refinements, including observation of temperature to the nearest tenth of a degree and the time precipitation had stopped or begun. But identifying the three-dimensional structure of the discontinuities remained a problem, especially without regular, direct aerological observations. They still needed some means of including weather conditions in higher strata of the atmosphere in the forecasting system.

If they could only delineate the discontinuities aloft, they could demonstrate their three-dimensional character and also help aviation, for abrupt changes in wind velocity, temperature, and pressure were known to be hazards for airships and for airplanes. Cloud observations used during the previous summer, and Margules's equations, had helped Jacob visualize the sloping surface of the steering surface. Furthermore, the summer forecasting service had used a number of prognostic signs derived from folklore for tracking cyclones that were still beyond the horizon. Herein lay a clue: by interpreting finely

29. *CIWY, 1918–1919*, p. 351. The terms *steerling lines* and *squall lines* were still used on occasion, but to refer to the line on the map where the boundary surfaces intersect the earth's surface.

30. Ibid.

distinguished cloud observations as signifiers of physical processes, they could reproduce surfaces of discontinuity as three-dimensional atmospheric structures.

The Sky as a System of Signs

If the three-dimensional physical boundaries between air masses were to be the primary conceptual focus for weather prediction, the clouds and weather along these boundaries could be used as part of the forecasting practice—a reversal of prewar forecasting practice. Although types of clouds had long been recognized as indicators of weather change, this folklore had no place in prewar forecasting, nor did scientific classifications of cloud forms that were developed during the nineteenth century.[31] Cloud forms entered forecasting work during the war, when aviation's needs forced forecasters to include them. The Bergen meteorologists found that clouds could also be interpreted as signifying the presence of three-dimensional surfaces of discontinuity. Indeed, an atmospheric semiology, based on finely differentiated weather phenomena, enabled the Bergen meteorologists to introduce surfaces of discontinuity into forecasting and to establish them as real scientific objects.

Using clouds as signifiers of atmospheric processes occurred to Bjerknes during the setting up of the Bergen forecasting service. While traveling up and down the coast in 1918, to the outermost reefs and to the remote ends of the fjords, he acquired new insights into weather prediction. From the farmers, lighthouse keepers, fishermen, and sailors who were to be observers for the forecasting service, he learned a rich folklore of weather prognostics.[32] Interwoven with ancient and superstitious beliefs were useful predictive signs. To help compensate for the missing data from the west, he realized, a system of signs might be devised that, when interpreted correctly, might provide clues to the weather systems approaching from beyond the North Sea horizon. Still on this journey, Bjerknes wrote to Arrhenius:

> I am traveling around the country erecting weather stations, setting up barometers, and speaking with weather-knowledgeable persons. . . . We can learn much from [sea]pilots and lighthouse keepers that cannot be

31. Several attempts to include basic cloud forms were rejected by the International Meteorological Committee. Ekholm, for example, recommended in 1891 that the international telegraphic code include the amount of cloud, distinguishing between lower and upper cloud forms; Air Ministry, *Third Meeting of the Commission on Weather Telegraphy* (London, 1921), p. 4.
32. V. Bjerknes to H. Bjerknes, 31 May 1918, BFC.

gotten at a desk. I am not without hope that their reports will be able to compensate the earlier telegrams from England. For a cyclone can be seen far out at sea, surely halfway over to England; that is, when seen by the right eyes, that do not confuse an innocent fog bank with the "dark air" ["*mørke luft*"] that forms a stratum over the cyclone at a height of 10,000 meters or more. This "dark air" is surely the cirrus layer, and perhaps yet higher cloud layers, *seen from the side*, while the thick rain clouds lie as yet far beneath the horizon . . . I came across this trail through visits to lighthouses and the navy's lookout stations.[33]

"Seeing" the right signs correctly could perhaps help compensate for the missing data.

Bjerknes then improvised instructions for the 1918 summer forecasting service's coastal observers. Systematic observations and reports of basic cloud forms, their orientation in the sky, and their direction were requested, both as a trial for future aviation forecasting and as an experiment for using prognostic signs.[34] Each observer was asked to scrutinize the entire sky visible to him for any telltale indicator of change in the weather. To help in this respect Bjerknes incorporated into the code several characteristic signs from folklore. The "dark air," or "brewing up" ("*optræk*"), noted in the letter, was one of the major prognostics: "Especially important is the observation of the 'brewing up' which is a sign that bad weather is approaching."[35] Once these prognostic signs were incorporated into the observation code, they were employed in the forecasting practice as a supplement to the wind-flow charts. In addition to signs of "brewing up," Jacob also used the direction cirrus clouds moved in his attempts to predict the motion of cyclones, following a finding Hesselberg had made before the war.[36] These observations were at that time, however, mainly a separate complementary aspect of a forecasting experiment based on lines of convergence.

In 1919, when the Bergen group began systematizing forecasting practices, the observations of clouds and the prognostic signs could be integrated into the new cyclone model (Fig. 15); as Bjerknes noted, "It is not astonishing that a natural phenomenon of such power may

33. V. Bjerknes to Arrhenius, 4 June 1918, SAP. "Mørke luft" was translated both as "dirty air" and as "dark air."

34. The poster that provided information on the weather service, "Veirvarsling under sommeren: Til opslag paa telegraf- og telefonstationer," and the special instructions for coastal weather observers, "Tillæg til veiledning for veiriagttagelser ved kyststationer," June 1918 and n.d. [June/July 1918] in VpVA. Hereafter cited as "Veirvarsling under sommeren" and "Tillæg."

35. "Tillæg."

36. J. Bjerknes, "Veiroversigter for Vestlandet sommeren 1918, I," and "Daglige veiroversigter sommeren 1918, II," JBP.

be noticed by different kinds of signs long before its arrival. Such signs are well known to fishermen and sailors. And the view to which we have thus arrived concerning the structure of cyclones is of particular interest in making us understand some of these signs and allowing us to utilise them for forecasting."[37] He then provided a detailed account of how the different cloud and rain formations could be seen as signs of the passing of a cyclone and its surfaces of discontinuity:

If we find ourselves north of the cyclonic path we have generally a dry and cold East wind. But some thousand metres above our heads we have the current of warm air flowing northwards. And this current is visible from below by the clouds formed in the warm air just above the surface which separates it from the cold air. First we see light cirrus, of a special structure, the first mark that a storm is brewing. In typical cases these cirrus clouds change gradually into a thin milky veil, through which sun and moon are seen as pale discs, often surrounded by halos, which, according to the forecasters of the people, give still stronger evidence that a storm is approaching. The veil gets gradually denser, changing into alto-stratus. Then single raindrops begin to fall, gradually increasing to continuous rain, which by and by gets denser before gradually clearing again; the clearing is followed by a gradual change of the direction of the wind from East to North. At great distances north of the centre only a bank of cirro-stratus is seen to pass slowly by in the south; and at still greater distances up to 1000 m. from the centre the clouds are no more recognized as cirrus, but seen as a narrow dark band near the horizon.

If we are south of the center the sky initially changes its appearance in the same way as in the preceding case. The precipitation begins in the same gradual manner, and continues as dense rain for a period which may extend from a few hours up to a half or a whole day. But then suddenly [with the passage of the steering surface (warm front)] the wind turns from SE to S or SW, and the temperature often rises several degrees. The cover of clouds changes from continuous strato-nimbus to fracto-nimbus, above which alto-cumulus may be seen. The rain stops and the weather remains in general fine, though a heavy shower-cloud may now and then pass by. But then, after a period varying from a few hours to a day or two of fine or relatively fine weather, a sudden change comes [as the squall surface (cold front) approaches]. As a first warning generally a front of alto-cumulus is seen to approach, from the horizon, soon followed by heavy rain-clouds. The rain begins suddenly with full intensity, in some cases accompanied by thunder, but always with a sudden turning of the wind from SW to NW and often a very striking fall of temperature. But after a few heavy showers the final clearing up follows,

37. V. Bjerknes, "Structure of the Atmosphere," p. 129; see also in shorter form in "Forutsigelse av regn," p. 334.

provided that the rain does not continue by reason of local causes, as is often the case on the west coast of Norway.[38]

This account should be understood as the beginning of a system of signs that the Bergen meteorologists continued during the next few years to elaborate into a powerful analytic method: an "indirect aerology," by which clouds and other weather signs compensated for the lack of direct observations of the upper air.[39] They interpreted cloud observations to infer three-dimensional atmospheric structures and processes; for example, regarding the upper limits of the steering surface (warm front):

Even the tufted cirrus is a sign for the existence of a surface of discontinuity at that [extreme] height. The tufted cirrus must consist of ice crystals falling from the tufts through a well defined surface of discontinuity into a layer with different velocity from that of the tufts. . . . It is not improbable that we here observe the uppermost part of the warm current above the warm front surface. . . . Also the small cirro-cumulus ondulatus clouds, often forming in front of the veil of cirro-stratus, seem to indicate the existence of the same surface of discontinuity.[40]

A new epistemology of forecasting thus began to emerge. Knowledge of the atmosphere now came to reside in the interplay between map and sky, each informing the other, and thereby providing a means to picture three-dimensional atmospheric structures and predict change. Detailed observation of the sky became integrated for the first time into practice; the changing panorama of sky itself had become a source of knowing and part of a new method:

The appearance of the sky in the different parts of a cyclone, at different distances from its centre and in different situations relatively [sic] to the steering and squall lines, is so characteristic, and develops in so typical a way during the passage of the whole system, that it will always be recognized when one has once become acquainted with it. . . . Observation of the phenomena of the sky will be seen to have an importance equal to the study of the weather chart, especially for short-range forecasts. The time should be past when weather forecasts are made as bureau work in an office from which only [a] narrow strip of the sky is seen.[41]

38. Ibid., p. 130; in Norwegian, pp. 334–35.
39. Tor Bergeron, who joined the group in 1919, coined the phrase and did most to perfect the method.
40. J. Bjerknes and H. Solberg, "Meteorological Conditions for the Formation of Rain," *GP* 2, no. 3 (1921), 27.
41. V. Bjerknes, "Structure of the Atmosphere," p. 130.

Bjerknes's insistence that the sky be included in the preparation of daily weather forecasts did indeed represent a major transformation of meteorological practice. Not only would the observers around the country report detailed weather and cloud phenomena, which would then be recorded and "read" on the map, but the forecasters themselves would also have to keep watch on the entire sky up to the moment they issued the forecasts. Weather bureaus could no longer be located in downtown buildings where the sky was obstructed. Forecasters must have access to the sky; in this manner the forecasters could

> keep abreast of the weather, partly with the help of instruments and partly by means of regarding the sky. On the basis of the new weather forecasting methods this is a very important factor. Previously, one could, to a certain degree, be content to study the weather maps; while [now] it is no longer possible to utilize fully the new methods without following for oneself the development of weather conditions all day long.[42]

> My assistants unanimously say, that they could no more think of performing forecasts at an office where they have not a good view of the sky, least of all if they should have to give the precise short-range forecasts of 6–12 hours, which aviators required.[43]

This forecasting strategy was ingenious. True, other meteorologists were beginning to look out the window as well, as the need to report cloud cover for flying became an accepted part of forecasting. But for most meteorologists the opening of the sky only meant a more elaborate system of empirical, formal rules for prediction; clouds were simply added to the other observations. They sought no deeper insight into physical processes while engaged in the forecasting work. In the Bergen system, in contrast, the changing patterns in the sky were to be read as signs for three-dimensional processes occurring along the surfaces of discontinuity. The new cyclone model, based on discontinuities, endowed the vastly enriched observations of weather and clouds with physical meaning and made it possible to forecast the occurrence of clouds and weather with a degree of localization unthinkable with other forecasting systems. At the same time, the ability to integrate surfaces of discontinuity into a meteorological system depended in part upon making and collecting all these finely dis-

42. V. Bjerknes to KUD, 8 December 1919.
43. V. Bjerknes to Shaw, 22 November 1919, copy, VBP.

tinguished observations of weather phenomena as part of forecasting operations; these were indispensable for reproducing the discontinuities in practice. As he had done earlier when first considering an atmospheric physics, Bjerknes tried once again to assess the resources available and the manner by which the atmosphere was to be used, and once again he began developing a system with which he might create a stable national discipline and dominate meteorology internationally.

The Historical Specificity of the Bergen Cyclone Model

The Bergen meteorologists developed a nonevolving cyclone model based on surfaces of discontinuity (or fronts) as part of their efforts to establish a new system of weather forecasting. Bjerknes and other meteorologists had long recognized the existence of various forms of atmospheric discontinuities; the notion of a discontinuity, or rapid transition zone, was not new. Theoretical calculations showing the possibility of discontinuities and prior descriptions of weather phenomena associated with the Bergen model imply that the rationale for conceptualizing such structures and the conditions for integrating them into meteorology first came about after the start of World War I. Indeed the changes in technological and observational bases for forecasting, and changes in the goals of forecasting—specificity and detail—were preconditions for the Bergen cyclone model. By focusing on surfaces of discontinuity, the Bergen meteorologists could attempt to localize weather phenomena in time and space; the demand for and the technological-economic possibility of attaining such localizations only emerged during World War I. The significance of the Bergen model can be more easily understood in the context of some earlier cyclone models and forecasting practices.

In the first edition of *Forecasting Weather* (1911) Shaw discusses a number of forecasting methods and cyclone models; they demonstrate how practice, perception, and conceptual development are interrelated. Shaw presents whole sections of the Hon. Ralph Abercromby's *Principles of Forecasting* (1883), in which many of the forecasting rules and the cyclone model used by the Meteorological Office were first established. Of the many storms analyzed, that of 12–14 November 1875 received special attention; this same storm was again analyzed in the 1930s by Tor Bergeron, who joined the Bergen group in 1919.

Abercromby drew a series of concentric circles in the field of pressure and derived a model for a typical cyclone based on this particular

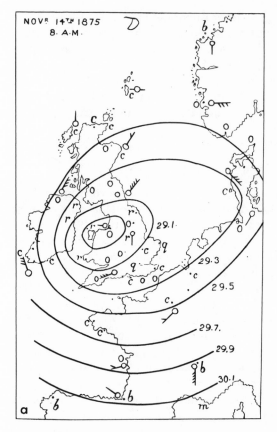

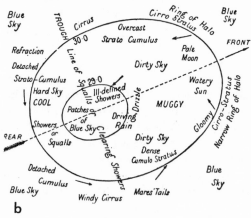

Fig. 16. Cyclone, 14–15 November 1875, Abercromby's analysis: (*a*) his weather map showing ovals in the pressure field; (*b*) his cyclone model. From William Napier Shaw, *Forecasting Weather* (London, 1911), pp. 87, 94.

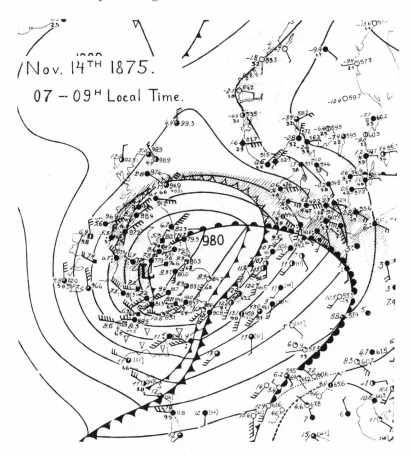

Fig. 17. Cyclone, 14–15 November 1875, Bergeron's analysis: He delineates a warm front over the Continent, furthest east; a cold front east of Britain; and an upper front (occlusion) connecting the rain over Denmark to that over Ireland. From Tor Bergeron, "L'utilisation des météorogrammes pour les recherches synoptiques," in *Procès-verbaux des séances de l'Association de météorologie*, vol. 2 (Lisbon: 1933), p. 11.

storm (Fig. 16). Some fifty years later Bergeron analyzed the same data.[44] Using a system of fronts, Bergeron connected the rain over northern Germany and Denmark, which had previously not been associated with that cyclone, with the rain over Britain and Ireland (Fig. 17). He analyzed the observations to delineate a distinct warm sector bounded by fronts (east of Britain). When he read the "metro-

44. Tor Bergeron, "L'utilisation des météorogrammes pour les recherches synoptiques," in *Mémoire présenté à l'Association de météorologie U.G.G.I.* (Lisbon, 1933), pp. 3–16.

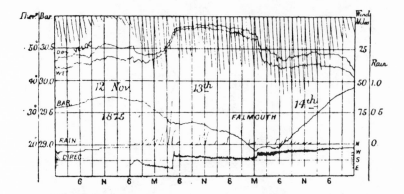

Fig. 18. Observations at Falmouth, 12–14 November 1875. A self-recording "metrogram" showing from top to bottom: wind speed (vertical lines), temperature and wet bulb temperature (graphs), air pressure (graph), rain (vertical lines), and wind direction (graph). From Shaw, *Forecasting Weather*, p. 93.

gram" from Falmouth, on which several weather variables had been automatically registered, he understood the rise in temperature and the shift of wind from southeast to southwest between midnight and 6:00 AM as a clear indication of the storm's warm front passing (Fig. 18). According to his interpretation, Falmouth then experienced the warm sector until roughly the following midnight. At that time, the storm's cold front passed, as indicated by the sharp drop in temperature, rise in pressure, and shift in wind from southwest to northwest. Abercromby, a former sea captain and keen observer of weather, had noticed these changes, had even described them, but continued to draw concentric ovals: the cyclone was for him—as for most forecasters for the next half century—virtually defined by the pressure field.[45]

> Out of the study of synoptic charts there arose immediately a necessity for the classification of two characteristic types of closed isobars—those which were associated mainly with strong winds, and represented a circulation of air . . . round a low pressure centre, counterclockwise, and

45. F. H. Ludlum, *The Cyclone Problem: A History of Models of the Cyclonic Storm* (London, 1966), discusses Bergeron's analysis and notes that "the same 'facts'" that fitted the earlier model are interpreted "in an elegant and convincing manner" by Bergeron. Ludlum's sensitive analysis of how differing cyclone models have been based on the same facts concludes that meteorologists differed in what they thought was "significant." I contend the notion of significant is largely historically contingent and closely related to goals and methods of forecasting practice.

those which were concerned mainly with light winds, and represented a gentle circulation of air clockwise round a centre of high pressure.[46]

When a system of isobaric lines has been drawn we get a picture of the pressure distribution over the region represented. The universal, in fact the necessary, practice is to single out and mark certain particular regions LOW and HIGH [pressure].[47]

Shaw presents, and in part organizes his book around, a natural history of pressure configurations, a classification of forms of the isobars. The lines of surface pressure are classified by shape: cyclone, anticyclone, straight isobars, secondary depressions, V-shaped depression, wedge-shaped depression, and col. "We can formulate the results of prolonged study of weather maps by defining types of weather, or, more strictly, types of barometric distribution, draw conclusions as to their relationship, and hence, as to the sequence of types,"[48] he writes. In his discussion of a V-shaped depression, in which the isobars form concentric parabolas with the lowest pressure in the innermost parabola, Shaw describes what the Bergen meteorologists would say was the passage of a steering surface (warm front):

The V shows very clearly the invasion of the region of easterly wind by a southerly wind with a marked difference of temperature, and rain or snow over the junction. The 40°[F] isotherm [line connecting points having the same temperature], which has been drawn on the chart, marks the separation of the warm and cold currents. . . . I was in Birmingham. . . . The evening was bitterly cold with a penetrating east wind. At 10 o'clock it began to snow, and much snow fell between 10 o'clock and midnight; shortly after midnight the southerly wind displaced the easterly at the surface, and rain with a sudden thaw supervened . . . such sudden changes are characteristic of V-shaped depressions.[49]

He then describes the passage on the following day of a second V-shaped depression, but of "opposite character": the axis running north–south (over Britain) instead of west–east, cold air is replacing warm air as the trough of lowest pressure passes. A warm sector can be imagined between this V (squall surface/cold front) and the earlier

46. Shaw, *Forecasting Weather*, p. 22.
47. Charles F. Marvin, "Introduction," in U.S. Dept. of Agriculture, Weather Bureau, *Weather Forecasting in the United States* (Washington, 1916), p. 25.
48. Shaw, *Forecasting Weather*, p. 51.
49. Ibid., pp. 65, 68–69.

one, now over Denmark. Many such examples can be listed, including examples from Bjerknes and his assistants' pre-Bergen studies in which "warm waves" and "cold waves" as well as lines of convergence at sea reveal many of the structural features of the 1919 steering and squall surfaces.[50]

Shaw, in fact, arrived at a cyclone model from an innovative study of trajectories of air currents that reveals many of the structural features of the Bergen 1919 cyclone (Fig. 19). In addition to the obvious asymmetrical character of the cyclone, Shaw described abrupt changes in wind, temperature, and pressure where one type of air displaces another. He did not, however, conceive of any structure or concept marking the transition from one air current to another. When Bjerknes and his assistants learned of this model in 1919 (probably from Bergeron), they diplomatically gave Shaw credit as an independent precursor.[51] Bjerknes noted that Shaw's model had not had any impact on meteorologists' conception of a cyclone; certainly not in forecasting work—not even at Shaw's Met Office in London.

Focusing on the structural similarities actually misses the innovation of the Bergen achievement, which was not the picture of a cyclone with a warm sector; rather, it was the establishment of the surfaces of discontinuity as entities that constitute cyclones and that underlie a forecasting practice. The Bergen school's accomplishment in its early years entailed a transformation in practice that subsequently produced a transformation in theory. Abercromby, Shaw, and many others had noticed what might be called frontal phenomena but did not make them the object of study or conceptual development. Studies of lines squalls treated this phenomenon as autonomous and not part of cyclones. Even after the theoretical conditions for noting the existence of atmospheric discontinuities had been provided by Helmholtz and Margules, occasional observations of sharp transitions were considered merely the displacement of one air current by another. The boundary between the air currents first became—and could become—a conceptual entity, the focus of practice, and the basis of a cyclone model when the character of weather forecasting changed after 1914.

Without dense networks of observation stations and elaborate, rapid communication systems, surfaces of discontinuity would be meaningless in forecasting. The amount of data used in Bergeron's analysis of the 1875 storm could not have been included in nineteenth-century forecasting operations. Although these data were recorded, only a small fraction of them were telegraphed to a central

50. *Dynamic Meteorology and Hydrography*, 2:60–61.
51. V. Bjerknes to Shaw, 21 March 1919, copy, VBP.

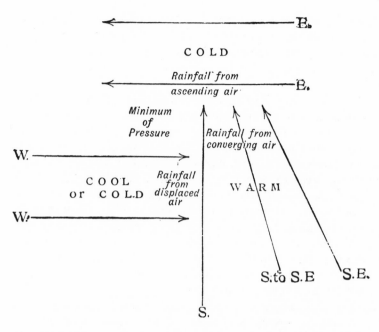

Fig. 19. Shaw's cyclone model (1906/1911). Shaw and R. G. K. Lempfert first published this model in *The Life History of Surface Air-Currents: A Study of the Surface Trajectories of Moving Air* (1906). From Shaw, *Forecasting Weather*, p. 212.

forecasting office. The telegraph systems of the time could not transmit observations from so many stations quickly and cheaply enough. The network of stations that actually did send telegrams to London, Paris, Christiania, or wherever was much too sparse for forecasters to consider including discontinuities or sharp transition zones in forecasting. Shaw observed that strong variations of pressure found on a chart from a self-recording barometer at a given observing station will often not be represented on the synchronous charts that show the geographic distribution of pressure at a given time. The reason, he explained, was that "the charts are prepared from observations at places a great distance apart, and the tendency is to round off or smooth the isobars. . . . Sudden changes of wind are often related to equally sudden changes in the direction of isobars, but as a rule the changes are smothered out on the map, because, for one reason, there is generally no means of knowing the precise locality at which the sudden change occurs."[52]

52. Shaw, *Forecasting Weather*, p. 36.

The significance of discontinuities lay, moreover, in their contribution to forecasting with greater spatial and temporal precision; nineteenth-century telegraphy also precluded sending forecasts to specific locations for short, specific periods of time. Gale and storm warnings for coastal regions, especially for sailing ships, could protect ships and lives without being very specific about location and time. Because governments were not directly involved in commerce, assisting commercial ventures through weather forecasting was not yet part of political economics. Only minimal forecasting assistance to save lives, save insurance companies huge outlays for lost shipping vessels, and protect naval vessels belonging to the state could be justified. Otherwise, as Shaw pointed out, "For many days in the year the weather, good or bad, is taken by many people as it comes, with only the amount of grumbling that the art of conversation demands, and it is only on occasions of some exceptional importance that they indulge in the eighteen-penny luxury of [receiving] a forecast by telegram."[53]

By 1919 the technological possibility of creating a practice that focused on finer grained atmospheric structures arose as a consequence of the spread of telegraph and telephone networks and the introduction of wireless telegraphy. Observations from a dense network could be collected quickly enough to be useful in forecasting, and predictions for specific locations could be sent to many scattered places in time to be useful. When forecasters accepted the challenge to secure aviation and to promote the use of weather for efficient commercial and military activities, they provided the rationale for increasing the flow of weather information and focusing on finer grained atmospheric changes.

And of course Bjerknes and his assistants came to understand that with regular flying, analyzing and conceptualizing atmospheric systems in three-dimensions would be crucial for any new reformed meteorological system. Data obtained from balloon ascents, mountain observatories, and kites during the late nineteenth century enabled atmospheric scientists to begin considering a three-dimensional atmosphere. The steady increase in availability of upper-air observations before the war precipitated statistical investigations correlating temperature and pressure patterns in the upper air with patterns near the surface. Meteorologists and aerologists sketched gross features of the vertical and horizontal distribution of air currents in high- and low-pressure systems. But neither they nor Bjerknes and his assistants in the pre-Bergen studies defined specific three-dimensional structures that might account for the observations. During and after the

53. Ibid., p. 368.

war meteorologists recognized that their models and practices could provide only limited information about where sudden shifts of wind, temperature, or pressure might occur, or where and at what heights icing on aircraft or a low-ceiling might be met. Similarly, after 1914 meteorologists accepted the need to differentiate fine-scaled differences in types of weather and to introduce cloud formations into their codes. These changes in classification, and therefore in perception, enabled the Bergen meteorologists to establish a technique for reproducing three-dimensional discontinuities in forecasting work.

Just as the changes in the technological basis and the goals of forecasting did not cause the Bergen cyclone model's creation, neither did the postwar need to comprehend weather phenomena in three-dimensions. All the changes in means and aims, along with theoretical resources, simply established a grid of possibilities for articulating new concepts; that is, they regulated in part some of the underlying criteria for producing knowledge that could be useful and meaningful in a meteorological system. They helped determine what scale of phenomena ought and could be conceptualized, what types of models of atmospheric structures could be fruitful, and what degree of exactness was acceptable. Practice, to a very large extent, regulated the development of concepts and the interpretation of observations. Just as a concept must rely on existing theory and a structure of ideas to endow it with meaning, so it can function in relation to other assumptions and principles, so it must also be integrated into a practice if it is to be reproducible through a set of analytic techniques. Theory and practice do not stand in opposition to one another or separate from one another; they are interconnected and interdependent. Analytic techniques—be they mathematical or instrumental—establish the objects of scientific study.

I am not claiming that knowledge simply fits functionally into a set of criteria; obviously the postwar possibilities and challenges did not force all meteorologists to think alike. Bjerknes, with his extraordinary talent for recognizing the social currents sweeping meteorology along, understood the opportunities in postwar Europe. In contrast to Shaw's innovative prewar study of the trajectories of air currents, which resulted in a new depiction of the cyclone but no new forecasting practices, the Bergen meteorologists forged their model together with predictive methods. Jacob's initial discovery was a new phenomena explained and depicted largely in a preexisting mode. In 1919, after a series of modifications on the original discovery, the Bergen group effected a transformation of the explanatory scheme, as they simultaneously set forth a new way of forecasting weather and a new cyclone model.

Changes in practice prompted and enabled the Bergen meteorologists to introduce surfaces of discontinuity into meteorology as a
central concept. The emergence of this concept entailed new criteria
for seeing and using weather. The opening of the sky as a system of
signs within meteorological discourse, with the accompanying need to
change the architecture of weather bureaus (from inclosed downtown
offices to towers that permitted direct viewing of the sky) is part of the
same history. In turn this entire development cannot be divorced
from a fundamental event: the advent of regular commercial and
military flying:

> It was [long ago] learned that a rise of air-pressure often meant good
> weather and that when the air-pressure fell, bad weather might be ex-
> pected. Forecasting procedure depended solely upon the treatment of
> surface data and the relation of moving air-pressure systems to weather
> changes; no serious attempt was made to offer an explanation of weath-
> er phenomena based on sound physical principles. With the develop-
> ment of commercial and military aviation, a better method became im-
> perative. . . . The need for a study of weather from a three-dimensional
> instead of a two-dimensional point of view became apparent.[54]

Bjerknes well understood these changes as they unfolded. He and his
assistants still had much to accomplish.

54. U.S. War Dept., *Technical Manual No. 1-230: Weather Manual for Pilots* (Washington, D.C., 1940), pp. 3–4.

[9]

The Bergen School's Pursuit of International Leadership: Constituting the Polar Front

PASSIVITY was never part of Bjerknes's strategy for achieving professional success. After the war he gradually resumed his vigorous efforts to spread new ideas and techniques. Concurrent with the effort to define a weather forecasting system, he also began to strive for a Bergen school that could dominate international meteorology. During the fall of 1919, Bjerknes visited Paris and London for the twin purposes of presenting the nascent Bergen meteorology and of learning greater details about what measures would be necessary for a comprehensive weather service for aviation. Shortly thereafter the polar front was constituted in Bergen. As represented in the Bergen school's first accounts, the polar front was a single surface of discontinuity stretching around the northern hemisphere, separating polar air from equatorial air.[1] This concept was not simply "discovered"; it was first postulated. The rationale for constituting the polar front and the claims made for this concept (some of them exaggerated) arose from Bjerknes's desire to establish a forecasting system for aviation that could enable him and his assistants to achieve dominance in international meteorology. Efforts by Bergen meteorologists to disseminate their new meteorological system affected their own subsequent attempts to develop additional atmospheric models.

1. V. Bjerknes, "Om vær- og stormvarslinger og veien til at forbedre dem," *Teknisk Ukeblad* 38 (1920), 300–7; idem, "The Meteorology of the Temperate Zone and the General Circulation of the Atmosphere," *Nature* 105 (24 June 1920), 522–24, also pub. in *MWR* 49 (1921), 1–3. Subsequent page references are to *Nature*. The frequently cited article by J. Bjerknes and H. Solberg, "Life Cycle of Cyclones and the Polar Front Theory of Atmospheric Circulation," *GP* 3, no. 1 (1922), 1–8, contains many substantial modifications of the original formulation of the polar front.

Already in the fall 1918 Bjerknes had begun spreading information abroad about the preliminary Bergen findings and taking measures to establish a school of disciples.[2] He had quickly sent a report on Jacob's preliminary findings, together with notes on how these results could be useful in the United States, to Washington for publication in the *Monthly Weather Review*.[3] Closer to home, he had sent Jacob and Solberg to Sweden, where they lectured and recruited young scientists to come to Bergen to join their effort.

Young Swedish assistants could be important, both to help in the arduous forecasting work and to spread methods and research problems back home. Preliminary response from the elder Swedish meteorologists was not favorable; the Bergen methods were immediately dismissed as "humbug."[4] Of greater immediate concern was obtaining assistance; Jacob and Solberg had had to struggle to sustain the 1918 summer forecasting service. With the boom in the Norwegian economy, many university graduates with science degrees were taking high-paying jobs in private industry; others accepted teaching jobs even before finishing their education; the salary level for forecasters was low. The one Norwegian science student Bjerknes managed to recruit in 1919, Svein Rosseland, found the work uninteresting and soon left to pursue what became a highly successful career in astrophysics. Only in 1921, when the economy entered a depression, did they manage to attract another Norwegian, Finn Spinnangr, to the weather service. The many young Swedish scientists who came to Bergen, including Bergeron, Eric Björkdal, Ernst Calwagen, and Carl-Gustaf Rossby, proved instrumental in sustaining and developing the Bergen weather service and meteorology.

In late spring and summer 1919, Bjerknes invited young Danes, Norwegians, and Swedes to a course in methods. He also turned to Leipzig, hoping to teach Wenger, his former assistant and now successor, the latest developments. A stipend from the Norwegian commercial airline company enabled Wenger to join the course. He came skeptical; he left convinced a new epoch in forecasting was beginning in Bergen.

In the fall of 1919, Bjerknes turned to the new centers for international meteorology, Paris and London. He attended to Paris from 30

2. V. Bjerknes to R. S. Woodward, 9 August 1918; V. Bjerknes to Svante Arrhenius, 3 November 1918; V. Bjerknes to Minister P. Benjamin Vogt (Norwegian Legation in London), 17 February 1919; copies, all in VBP; J. Bjerknes to Arrhenius, 25 October 1918, SAP; Tor Bergeron to J. Bjerknes, 23 February 1919, UBO.

3. V. Bjerknes, "Weather Forecasting" and "Possible Improvement in Weather Forecasting with Reference to the United States," *MWR* 47 (1918), 90–100.

4. Reported in J. W. Sandström to V. Bjerknes, 22 August, 22 September 1918, VBP.

September to 6 October the International Meteorological Committee's first full-scale meeting since the war.[5] Although representatives from the Central Powers were not invited, the primary issues taken up were those touching almost the entire international meteorological community: applications of meteorology to aviation, including the international exchange of weather observations and the use of wireless telegraphy for transmission of information. This meeting was held in conjunction with efforts to establish international agreement on those issues.

At the Paris Peace Conference of 1919 the French government proposed that the national delegations meeting to draw up peace treaties should work to create a uniform legal framework for international air traffic. The resulting Convention Relating to the Regulation of Aerial Navigation established a basis for agreement on diverse aspects of aviation ranging from the sovereignty of each state over the air space above its territory to aircraft markings and standards for airworthiness.[6] Among the issues covered in the convention are those pertaining to civil aviation's meteorological requirements. Based on the immediate postwar understanding of these requirements, Article 35 and Annex G of the convention established guidelines for international cooperation between nations on the collection and dissemination of meteorological information and provided new codes specifying the type and detail of data to be obtained when making observations. The convention called for implementing a variety of forecasts covering different time scales, and to realize these new goals, it stipulated exchanges of observational data between nations several times daily by wireless telegraphy. Annex G also proposed a classification of one hundred varieties of weather phenomena as well as classifications for low-, medium-, and high-level cloud forms.[7] These proposals had to be ratified formally by the International Meteorological Committee.

Although the sessions at the meteorological congress were to be limited to organizational discussions and negotiations, especially related to the convention, one formal scientific lecture was presented during the seven-day meeting; it was delivered by Bjerknes. From the

5. Bureau Central Météorologique de France, *Procès-verbaux des séances de la conférence météorologique internationale des directeurs et du comité météorologique international réunion de Paris 1919* (Paris, 1919).

6. *Convention portant réglementation de la navigation aerienne (13 October 1919) Convention for the Regulation of Aerial Navigation (13 October 1919)* (London, 1920), published by the British Air Ministry in both languages. All page numbers refer to a French and English text facing each other. A copy of the convention was found in VBP; a Norwegian translation can be found in NMI.

7. Ibid., pp. 12, 35–37.

very start of the meeting, disagreements and confrontations developed, especially between the meteorologists and the military officers and aviators. Even Bjerknes, who was accustomed to the presence of military and aviation officials at the international aerological meetings before the war, expressed surprise over the large number present at the Paris meeting, even without German representatives. In an attempt to "cool down the pilots' anger" Shaw, president of the international organization, allowed Bjerknes to deliver a lecture. After presenting an account of the initial work of the Bergen school, which emphasized the use of atmospheric discontinuities for attaining accurate, detailed short-term weather forecasts, Bjerknes received the first applause of the conference.[8]

By the end of the meeting, when the International Committee's various specialized commissions were reestablished, Bjerknes was elected head of the Commission for the Exploration of the Upper Atmosphere (as the Commission for Scientific Aeronautics was now called) and was elected to the steering committee for the new commission on the applications of meteorology to aerial navigation. The new commission on applying meteorology to aviation would ascertain aviation's meteorological needs, including those relating to theoretical research and to transmission of weather information.[9] The aerological commission would ostensibly continue with the scientific study of the upper levels of the atmosphere and with the technological problems of investigating the "free air." In an unusual burst of emotion, Bjerknes wrote jubilantly to Honoria, "I am the new Hergesell!"—referring to the prewar president of the commission who had wielded so much influence in aerology.[10] He immediately sought and gained approval for one of his prewar proposals: the introduction of millibars as the accepted unit of atmospheric pressure.

After further discussions with French meteorologists, Bjerknes left for London, where he presented the Bergen methods and models, which he believed would "be of great importance for the coming air traffic," to British aviation officials and meteorologists.[11] Bjerknes was not only propagandizing on this trip; he was also attempting to learn more about the British and French understanding of aviation

8. V. Bjerknes to H. Bjerknes, 1 October 1919, BFC; Bureau Central Météorologique, *Procés-verbaux*, p. 17.

9. Bureau Central Météorologique, *Procés-Verbaux*, pp. 37, 51.

10. V. Bjerknes to H. Bjerknes, 10 October 1919, BFC.

11. V. Bjerknes quoted in Wilhelm Keilhau to G. Holt-Thomas, 4 June 1919, related to Norwegian-British aeronautical ventures; on Bjerknes's plans to speak with British aeronautical interests, see Keilhau to Bjerknes, 4, 25 June 1919; V. Bjerknes to F. Festing, 30 June 1919; copies, all in VpVA.

meteorology.[12] At the time of these trips he was preparing a report for the Norwegian government's Commission on Aerial Transport concerning what steps must be taken in Norway for securing commercial and military flying.[13] This formal inquiry gave Bjerknes an opportunity to study these issues systematically, especially in light of British and French perceptions of aviation's meteorological requirements.

The report also gave Bjerknes the opportunity to link his desire to increase the salary of weather forecasters to the needs of aviation. Only by increasing salary and providing time-off for research could the necessary scientific competence be attracted to the weather service. In fact, many improvements for the weather service—in technology, personnel, and observations—were needed to serve aviation, and Bjerknes prominently included them in the report to the national commission as well as in one to the Ministry of Church and Education.[14]

Once the Paris Peace Treaty was signed, Bjerknes and Hesselberg noted that if Norway was to join the proposed League of Nations it would have to fulfill the meteorological requirements stipulated in the convention. Correspondence reveals that in late November and December, Bjerknes and his assistants were working on modifying their forecasting practices in light Bjerknes's new information on aviation's needs. At this time too Bjerknes completed his detailed report to the Norwegian commission and sent it to Hesselberg for review. In mid-December Hesselberg stated that they had all been working toward "a comprehensive plan for the development of a weather service for aerial transport."[15] The Bergen meteorologists then postulated the concept of the polar front.

Two issues prominent that fall had proven crucial for the polar front's emergence: long-term forecasting giving general weather outlooks three to four days in advance and rapid hemispheric exchanges of weather information by wireless telegraphy. Both were made

12. V. Bjerknes to Lufttrafikkommisjonen, 8 August 1919, quoted in Lufttrafikkommisjonen to Theodore Hesselberg, 20 August 1919, NMI.

13. Oberst-Col. Gustaf Grünner, chairman of the commission, to V. Bjerknes, 29 July 1919, VpVA; *Instilling fra Lufttrafikkommisjonen* (Christiania, 1921), p. 3.

14. V. Bjerknes, "Betænkningen om den norske veirvarsling og dens utvikling," 1 September 1919, "KUD/D, folder: "Værvarslinga på Vestlandet," file: "Budsjettsaker 1916–1946"; V. Bjerknes, untitled report to Luftrafikkommisjonen, December 1919, VBP and NMI; letters from Hesselberg and V. Bjerknes to KUD, November and December 1919, "VpV"/Budsjettsaker."

15. V. Bjerknes to Hesselberg, 6 December 1919, NMI; Hesselberg to Det Norske Utvalg til Forberedelse av den 3. Interskandinaviske Luftfartkonference, 5 December 1919, letter book 26, NMI.

provisions of the Paris Peace Treaty's convention and incorporated in Bjerknes's report to the Norwegian commission. They had become prominent in the summer 1919 with the first successful transatlantic flights. After two airplanes flew east from Newfoundland and an airship made a round trip from Great Britain to New York, "Atlantic fever" raged on both sides of the ocean; commentators predicted that fleets of Zeppelin airships would soon traverse the ocean regularly. The extensive meteorological activities coordinated to provide weather information for safeguarding these first flights revealed the challenge of serving transatlantic flight was an awesome one. Transatlantic aviators had to contend with meteorological obstacles of a radically different order from anything previously confronting forecasters. Moving at high speeds with poor navigational equipment and often through vast areas of clouds and fog; being much more vulnerable than ships to sudden wind shifts, temperature and pressure changes, icing conditions, and severe weather; and having too limited a fuel capacity to allow for appreciable alterations in course, post–World War I aircraft could safely cross the ocean only when weather permitted. An aviator had to know in advance what weather conditions to expect while crossing the Atlantic and had to follow changing conditions in flight. Little wonder the weather over the North Atlantic and issues of forecasting for aviators crossing the ocean were prominent topics in meteorological literature and at conferences.[16]

These issues of course reinforced the Bergen meteorologists' interest in rapid exchanges of weather information. From the start in Bergen, Bjerknes's vision of a transformed system of forecasting entailed both conceptual and communications innovation, the two being interdependent. Then he realized cyclonic discontinuities could only be useful in forecasting if they could be traced well beyond Norway's boundaries from observations collected by wireless communications.[17] In September, Bjerknes claimed that the introduction of wireless telegraphic exchanges of weather data in some European nations foreshadowed major changes in the nature of weather prediction. He expressed the hope that these exchanges of data would eventually

16. See Richard K. Smith, *First Across: The U.S. Navy's Transatlantic Flight of 1919* (Annapolis, Md., 1974), pp. 11–13, 151, 191–92, 195, 255–57; and among many articles, Charles Franklin Brooks, "Meteorological Aspects of Transatlantic Flight," *Science*, 25 July 1919, pp. 91–93; Willis Ray Gregg, "Trans-Atlantic Flight from the Meteorologist's Point of View," *Aviation* 6 (1919) 370–72, 422–25; Robert De Courcy Ward, "Meteorology and the Trans-Atlantic Flight," *Science*, 1 August 1919, pp. 114–15; on "Atlantic fever" and general enormous interest in aviation in Norway in 1919, Per Vogt, *Jerntid og Jobbetid: En skildring av Norge under verdenskrigen* (Oslo, 1938), p. 207.

17. V. Bjerknes, "Om forutsigelse av regn," *Naturen* 43 (1919), 346.

expand so as to connect the European and American observation networks. Further improvement would accrue if the observations made from ships in the Atlantic Ocean could also be incorporated. Once accomplished, this transoceanic wireless network would provide an "excellent foundation" for the expansion of weather forecasting; moreover, these reforms in communications could be justified by aviation's need for "an accurate, fast, and reliable international weather service."[18]

Further thought on transoceanic communications was prompted by inquiries to the weather service from Bergen shipping interests. Mariners had begun to consider how to exploit the revolution in communications to transmit meteorological information to and from ships at sea:

> Negotiations are now in hand to extend and unify the system of collecting weather data by wireless from ships at sea all over the world, and at the same time to organize the free transmission of weather bulletins from a sufficient number of wireless stations . . . to provide weather reports and forecasts to ships at sea. . . . It should be borne in mind that the needs of ships at sea in regard to information regarding the weather will march hand in hand with the requirements of aircraft.[19]

The powerful Norwegian Shipowners' Association (Norges rederforbund) made inquiries regarding the expansion of the Bergen weather service to include in its sphere of operations the eastern Atlantic Ocean and the entire west coast of Europe. The federation, along with the Bergen Stock Exchange and Trade Committee (Bergens børs- og handelskomité), agreed that the value for the Norwegian economy, especially for shipping, fishery, and agriculture, justified the expense of obtaining weather data by wireless from the east coast of North American and various points in the Atlantic Ocean and Mediterranean Sea.[20] For shipping companies confronted with postwar competition and a drop in freight rates, "a weather service extended to the ocean will be a great advantage . . . good forecasts to ships on the sea making it possible for them for instance to save much coal."[21] The Bergen group, which already received handsome private

18. V. Bjerknes, "Betænkningen om den norske veirvarsling."

19. Admiralty Notice to Mariners, no. 880 (1919), "Wireless Meteorological Information to and from Ships at Sea" (London, 1919), copy, NMI.

20. Several exchanges of letters in September and November 1919 among the Geophysics Institute, Norwegian Shipowners' Association, and Bergen Stock Exchange and Trade Committee, together with a summary of the discussion (Bergens börs- og handelskomité to Geofysisk institut, 2 February 1920), VpVA.

21. V. Bjerknes to Charles F. Marvin (chief of the U.S. Weather Bureau), 1 August 1920, copy, VBP.

donations from shipping companies and shipowners, naturally paid attention to these requests that coincided with and reinforced its current lines of thought.

In December, Bjerknes wrote in the report to the Norwegian Commission on Aerial Transport that the geographic and quantitative expansion of the exchange of weather information to be expected because of aviation would have a major influence on forecasting: it might be possible to predict general weather outlooks several days in advance.[22] Thus both the goals and the technological possibilities for forecasting had broadened since the spring of 1919, when the Bergen team had focused only on short- and normal-term forecasting and cyclonic discontinuities. At the same time the conceptual foundation for forecasting was also broadened, as the empirical basis upon which the polar front was postulated was laid down.

The Polar Front as a Hemispheric Battleground

After the cessation of the summer 1919 forecasting service, Bjerknes and his assistants received authorization to issue, when necessary, storm warnings for the west coast during the fall and winter. Because both the frequency with which cyclones approach Norway's west coast and their intensity increases dramatically during the fall, the Bergen meteorologists obtained an excellent opportunity to refine their recent model and methods through daily analyses of the changing weather. Recent changes in European and Norwegian weather forecasting routine also helped. During the summer of 1919 some nations had begun to take observations four times daily; Norway had initially increased them to three times.[23] Not only did extra sets of observations provide meteorologists with further raw data for aviators; in the case of the Bergen meteorology, in which each map analysis was also an investigation, extra data helped expand the understanding of the development and movement of weather systems. For example, a cyclone that first appeared on a weather map when the morning observations were analyzed could, within twelve or twenty-four hours, pass beyond the map's boundaries or develop in an unexpectedly dramatic fashion. Each extra observation period, therefore, provided further opportunity to study the evolution of weather systems and discover new phenomena.

One such discovery was that a mature cyclone can "give birth" to a

22. V. Bjerknes, untitled report to Luftrafikkommisjonen.
23. NMIÅ *1919–1920*, pp. 19–21, 27–29, 35–36.

new cyclone.[24] Solberg had noted during the late summer a situation in which a residual squall surface that was stationary gradually seemed to turn into a steering surface and develop into a new cyclone. Experience from the fall forecasting made this phenomenon clearer. On the tail end of a squall surface, extending out from a mature cyclone, a small wavelike pattern forms; this marks the development of the new cyclone. As cyclones passed Norway's west coast frequently during that fall, the meteorologists soon recognized that not all these storms were autonomous entities; some were linked together. There were secondary cyclones growing out of and following the primary ones. New and surprising though this discovery was, we should not mistakenly read it as a step inevitably leading to the conceptualization of the polar front. Tor Bergeron, who had been recently recruited from Sweden, wrote long, detailed letters to Swedish colleagues in which he described the ongoing work in Bergen. He pointed out the new understanding that cyclones can form on the tail of squall surfaces belonging to mature cyclones but indicated no hint of broader ramification.[25] He described several instances in which British meteorologists simply called for unsettled weather under "a gradually extending depression [low pressure]" but for which the Bergen group recognized a distinct pattern: "the outer tail of the squall line as usual broke off [*avsnördes*] and developed into the little cyclone over the canal. . . . Each time the Englishmen are just as surprised: 'a secondary has appeared.'"[26] Bergeron's enthusiasm for Bjerknes's "attempt to make meteorology an exact science," is clear from these letters. Even at this point, his and the group's exhilaration over the realization they were developing new insights and methods, of which the rest of the meteorological world soon would hear, is apparent; but no mention of a polar front can be found.

In mid-December, during one of the daily informal discussions of the weather situation, the Bergen meteorologists first speculated that a single line—in fact a "battle line"—stretched around the northern hemisphere.[27] The group had earlier tried to describe their new

24. V. Bjerknes to Ernst Gold, 27 November 1919, copy, VpVA; see also V. Bjerknes to William Napier Shaw, 22 November 1919, copy, VBP. Expressions related to birth and death of cyclones, mother cyclones and cyclone families begin entering the Bergen school vocabulary in 1920.

25. Tor Bergeron to Oscar Edlund, 24 October, 26 November 1919; Bergeron to Sandström, 12 December 1919, draft; both in TBP.

26. Bergeron to Edlund, 24 October 1919; see also ibid., 26 November 1919; both in TBP.

27. Interviews with Tor Bergeron, February 1976, June 1976, Uppsala. Bergeron claimed that the discussion occurred around a desk; those present in addition to him-

cyclone model (Fig. 15) in terms of a battle along the two surfaces of discontinuity: "We have before us a struggle between a warm and a cold air current. The warm is victorious to the east of the centre. Here it rises up over the cold, and approaches in this way a step towards its goal, the pole. The cold air, which is pressed hard, escapes to the west, in order suddenly to make a sharp turn towards the south, and attacks the warm air in the flank: it penetrates under it as a cold West wind."[28] Now they extended this World War I image to hemispheric dimensions. Someone suggested that the polar air is the "enemy," initiating an attack toward the equator, while in response the warm equatorial air counterattacks with thrusts toward the pole. Between the two opposing types of air lies a battlefront that extends around the hemisphere, marking the polar air's furthest advances—hence the name *polar front.* Like battles along the front as one army or the other attempts to advance in war, atmospheric skirmishes were conceived as occurring along the polar front; cyclones form on the front, representing struggles between polar and equatorial air, each air mass attempting to advance into the other's territory.

During the next several months the Bergen group alternated use of the expression "polar front" with "battle line" (*kamplinje*) or "battlefront" (*kampfront*).[29] The word *front,* used as in "polar front," referred both to the actual three-dimensional boundary and the two-dimensional discontinuity plotted on the surface-level weather map, the latter representing where the boundary surface intersects the ground. Subsequently, the steering and squall surfaces were renamed warm and cold fronts. This change in terminology stemmed from confusion that arose over the interchanging of the expressions steering line or surface and squall line or surface.[30]

---

self were the two Bjerkneses, Solberg, and either Carl-Gustaf Rossby or Erik Björkdal. Bergeron remembers the meeting occurring just before he left Bergen; correspondence indicates he left for Sweden between 15 and 20 December. Located in the Bjerknes family residence, the weather service provided an informal atmosphere for frequent discussions, at least when the arduous forecasting chores were completed.

28. V. Bjerknes, "The Structure of the Atmosphere When Rain Is Falling," *QJRMS* 46 (1920), p. 127; he first used a battle analogy publicly in a lecture of 3 May 1919; see idem, "Om forutsigelse av regn," pp. 332–33.

29. Interviews with Bergeron. He claimed the group consciously derived the name "polar front" from the World War I notion of a front, for they first believed the cold air to be the initiator of activity; see the use of this analogy in H. Solberg to Hesselberg, 28 March 1920, NMI (Solberg's first account of the work), and in V. Bjerknes's early statements on the polar front, e.g., "Om den almindelige atmosfäriske cirkulation," lecture given in Copenhagen, October 1920, VBP.

30. J. Bjerknes and H. Solberg, "Meteorological Conditions for the Formation of Rain," *GP* 2, no. 3 (1922), 11–12, 25; idem, "Life Cycle of Cyclones and the Polar Front," p. 4. Sakuhei Fujiwara, who was in Bergen for most of 1921, recalls, "The

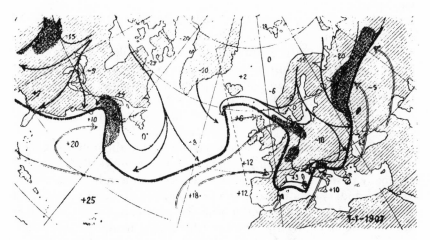

Fig. 20. Early polar front (1920). Solberg reanalyzed the weather situation for several days in January 1907 to show a single line of discontinuity stretching around much of the northern hemisphere. From V. Bjerknes, "Om vær- og stormvarslinger og veien til at forbedre dem," *Teknisk Ukeblad* 38 (1920), 306.

Once the idea of the polar front was suggested, Solberg looked for proof of its existence by reanalyzing a series of old weather charts that covered the north Atlantic Ocean and much of the North American continent. Danish and German meteorologists had constructed these charts from observations recorded in ships' logs and published sources; these provided one of the few available sources for analyzing weather conditions over an area larger than Western Europe or North America. According to the first statements on the polar front, these reanalyzed charts "showed" the existence of a major portion of the single circumpolar line of discontinuity (Fig. 20).[31] It has been suggested that Solberg "discovered" the polar front on these charts.[32] Yet these maps were based on observations from a sparse network of stations, and those from the oceanic areas were both scanty and poorly synchronized. Drawn as a narrow line of discontinuity separating two different types of air, the polar front could not be "dis-

words *warm front* and *cold front* were often used and gradually became fixed words" (p. 5) in his "The Men Who Had a Deep Insight into Nature: The Birth of the Norwegian School," trans. from the original Japanese article in *Astronomy and Meteorology* (August 1949) and privately mimeographed by the West Norway Weather Bureau (Vervarsling på Vestlandet) (1971). I thank Harald Johansen for providing me with a copy.

31. V. Bjerknes, "Om vær- og stormvarslinger," pp. 305–7; idem, "Meteorology of the Temperate Zone," pp. 523–24; Solberg to Hesselberg, 28 March 1920, NMI.

32. Theodore Hesselberg, "Utviklingen av vær- og stormvarslinger i Norge i de siste 25 år," *Vervarslinga på Vestlandet 25 år* (Bergen, 1944), p. 26.

covered," much less proven, from such charts. Even some members of the Bergen school recognized that this exercise merely depicted a hypothetical construct. Skeptical foreign meteorologists were totally unconvinced. When during conferences in Bergen and Leipzig in 1920, Jacob Bjerknes presented these maps and others on which he had drawn wavy polar fronts across the Atlantic Ocean, based on a few observations, he left many in his audience agape and unconvinced.[33]

Empirical evidence alone surely did not lead to the initial formulation of the polar front, a single surface of discontinuity. Rather, it was originally postulated at the conceptual level. Although the discovery that a secondary cyclone can form on a cold front extending out from an older cyclone was a prerequisite for articulating the concept, it does not explain why Bjerknes and his group made the jump to a hemispheric discontinuity, nor why this concept should become so important in their work.

The Polar Front and Forecasting Practice

The development of the polar front is more understandable in the context of the Bergen meteorologists' concern both with devising a forecasting system to satisfy the observational and predictive needs of aviation and with influencing international meteorology. The polar front was initially postulated when the battle analogy was extended from denoting a condition along the surfaces of discontinuity within individual cyclones to denote a single hemispheric "battle" front encompassing the individual skirmishes. The Bergen meteorologists at first believed that short-term forecasts could be provided by a forecasting practice based on these surfaces of discontinuity and grounded in rapid local communications. Now this entire system, extrapolated to hemispheric dimensions, could in principle also provide long-term forecasts. Belief that a hemispheric communications network would soon be implemented made it possible for the school to conceive of a hemispheric surface of discontinuity; indeed, the technology here seems to inform the concept. As the cyclonic discontinuities were originally conceived in the context of a rapid communications system, so the promise of a hemispheric network provided an impetus for expanding the notion of a surface of discontinuity to a similar scale, with the aid of the battle analogy. A single discontinuity, hemispheric in scope, could provide a conceptual focus for the needed circumpolar

33. Interviews with Bergeron; Bergeron to Solberg, 7 June 1924, HSP.

weather service, while the goals and materials of the weather service simultaneously could render the polar front a significant construct, as it could then be used to predict the weather.

During the winter Solberg and Vilhelm Bjerknes began elaborating the polar front into a central concept for both forecasting practice and theoretical study. They attributed the formation of both cyclones and anticyclones (high-pressure areas) to the motions of the polar front that followed attacks and counterattacks by polar and equatorial air.[34] They regarded the polar front, whose undulations swept across most of the temperate-zone lattitudes as the whole system moved from west to east, as a new means of comprehending the general circulation of the atmosphere and therefore of predicting the weather. Bjerknes and Solberg considered the meteorological events of the temperate zone to be "details" within the large-scale general circulation, which in turn were "correlated" with the motions of the polar front.[35]

In addition to underpinning practical forecasting, the polar front could also serve as an object of theoretical inquiry. Bjerknes foresaw returning to the problem of precalculating atmospheric changes by attempting to reproduce and predict the motions of the polar front mathematically. His earlier reflections that atmospheric disturbances arise from some form of wave motion could be developed into a mature theory by considering the polar front as just such a wave.[36] Indeed, later in 1920, Bjerknes started to dedicate his time to the elaboration of a polar front theory. Although he withdrew increasingly from daily weather forecasting activities, he still remained close to the overall forecasting endeavor, especially in his role of manager-entrepreneur-propagandist for the Bergen school. In this regard he appreciated that "the main thing is the result that the rational basis for the weather forecasts of the temperate zone is the survey of this 'polar front', which sweeps across this entire zone."[37] And in order to realize this predictive function, the polar front would have to be used in conjunction with a "circumpolar weather service." All initial refer-

34. Solberg to Hesselberg, 28 March 1920, NMI; V. Bjerknes, "Om vær- og storm-varslinger," pp. 303–7.
35. *CIWY 1920*, p. 389 (copy of TS shows the report was sent 2 July 1920); idem, "Meteorology of the Temperate Zone," p. 524.
36. V. Bjerknes, "Wellenbewegungen in Kompressiblen schweren Flüssigkeiten: Erste Mitteilung," *Abhandlungen SGW* 35, no. 2 (1916), 33–65; idem, "Theoretisch-meteorologische Mitteilungen," *MZ* 32 (1915), 337–43; idem, "On the Dynamics of the Circular Vortex with Applications to the Atmosphere and Atmospheric Vortex Motion," *GP* 2, no. 4 (1921), 1–88.
37. V. Bjerknes to Woodward, 20 April 1920, copy, VBP.

ences to the polar front are linked to discussions of such a service; together they form a single focus of attention—and they also reveal Bjerknes's desire to advance the Bergen meteorology.

Bjerknes turned first to Washington and London; he presented a plan for a circumpolar weather service. If a network of observation stations could be established stretching around the hemisphere and linked by appropriate communications facilities, it would then be possible to keep track of the polar front and its motions. A circumpolar weather service of this kind, based on the polar front, "would certainly be a great benefit to all occupations dependent upon the weather, such as agriculture, fishing, and shipping, and perhaps no less then a necessity for the realization of transoceanic air routes." Bjerknes claimed that predictions based on the polar front could provide precise short-term forecasts as well as general long-term ones. The polar front "can not fail to exert a considerable influence upon the methods of weather forecasting. . . . An effective survey of this front all round the pole will form the rational basis for short-range as well as of long-range weather forecasts." "These two kinds of forecasts could be extended to all regions of the temperate zone—oceanic as well as continental."[38]

Bjerknes's scheme for a circumpolar weather service consisted in uniting the various observation networks in the northern hemisphere nations and in supplementing these with a number of floating observation stations on ships and with several stations in the polar region erected for this specific purpose.[39] This aspect of Bjerknes's plan differed little from other designs for a hemispheric weather network except in one respect: he suggested "an international [weather] central" to use the hemispheric data to analyze the polar front and its motions daily and transmit the results to national weather services. The latter could go about their own regional duties, only with the addition of a "full knowledge of the general weather conditions of the whole northern hemisphere" and expected changes based on the study of the polar front. Thus this information could provide the basis for predictions of "the general character of the weather for much more than four days ahead, and at the same time give to the . . . forecasts for a day or two a degree of precision hitherto by far not attained."[40]

And where but Bergen was the expertise to analyze the polar front? Even if it proved impossible to locate the international weather central in Bergen, the central would still have to be staffed with persons

38. *CIWY 1920*, pp. 388, 389; idem, "Meteorology of the Temperate Zone," p. 524.
39. V. Bjerknes to Woodward, 20 April 1920; V. Bjerknes to Shaw, 1 May 1920, copy; both in VBP; V. Bjerknes, "Om vær- og stormvarslinger," pp. 7–8.
40. V. Bjerknes to Woodward, 20 April 1920, VBP.

well-trained in the Bergen methods. According to Bjerknes, at least ten meteorologists would be needed to analyze the polar front and to teach others how to use these techniques: "But of course these meteorologists had to be so completely trained in the analysis of the very detailed meteorological charts, which have given the key to the secret."[41] Secrets imply possession; in this instance, the "key" entailed the analytic techniques for reproducing the polar front and related cyclonic fronts in forecasting practice. The Bergen school's "microscopic analysis," which stressed a physical interpretation of each observation, had to be learned from someone possessing this skill.[42] To learn how to use this system of forecasting, meteorologists would have to come to Bergen, or invite Bergen scientists to their universities and weather forecasting institutions.

The polar front model was initially informed in part by the role it was meant to serve in forecasting. Cyclones were said to develop and move as part of the polar front—"like pearls along a string"; hence the cyclonic precipitation, cloud cover, wind, and temperature structure specified in the 1919 Bergen model could be forecast from a survey of the polar front. In this manner, short-term and "normal" forecasts could be deduced. Although a mechanism by which cyclones evolve was not as yet clear, the Bergen meteorologists conceived cyclones as some form of wave moving on the polar front. Along those portions of the polar front that do not constitute cyclones, fog occurs. An idealized section of the polar front was represented by the diagram in Fig. 21 and described as follows: "Along the whole of this [polar] front line we have the conditions, especially the contrasts, from which atmospheric events originate—the strongest winds, the most violent shifts in wind, and the greatest contrasts in temperature and humidity. Along the whole of the line formation of fog, clouds, and precipitation is going on, fog prevailing where the line is stationary, clouds and precipitation where it is moving."[43] Immediately apparent in this account is the reference to the weather phenomena most important for aviation, especially transatlantic flight; strong wind, especially during a pronounced shift in wind direction, and sharp temperature and humidity changes were recognized problems for airships and airplanes.[44] Although Bjerknes's remarks imply concern with these problems, what is most remarkable is his attribution of

41. V. Bjerknes to Shaw, 1 May 1920, copy, VBP.

42. Bjerknes frequently called the Bergen methods "precision techniques" or "instruments"; see, e.g., his "Om den almindelige atmosfäriske cirkulation," VBP.

43. V. Bjerknes, "Meteorology of the Temperate Zone," p. 524.

44. Among the relevant articles, C. LeRoy Meisinger, "The Weather Factor in Aeronautics, *MWR* 49 (1920), 706.

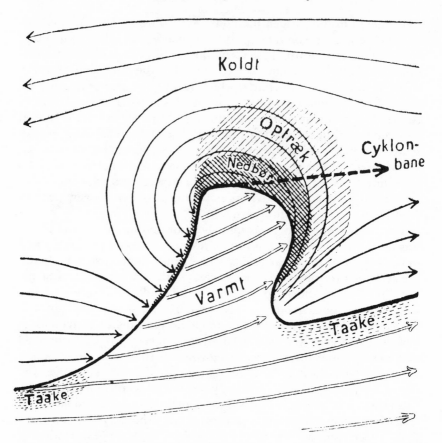

Fig. 21. Cyclone along a polar front (1920). Taake = fog; Nedbør = precipitation; Optræk = "brewing up"; Cyklonbane = path of the cyclone. From V. Bjerknes, "Om vær- og stormvarslinger," p. 301.

fog to the "stationary" portions of the polar front. All the literature written of the time on the problem of transatlantic flight and "oceanic meteorology" discussed the hazards posed by the huge stretches of fog over the Atlantic Ocean, which could not be reliably predicted by then-prevalent methods. Bjerknes and his colleagues endowed their new construct with the ability to account for these fogs and, as a consequence, with the means of predicting them.

In retrospect such claims can be seen as exaggerations. Yet it is the very cavalierness with which Bjerknes and his assistants constructed the first model of a polar front that reveals their concerns. Why would they definitively characterize the polar front by these regions of fog, when they could offer no physical reason why fog must occur where the polar front is stationary? This characteristic was in fact soon dropped, in subsequent articles on the polar front. The idea of a *single* discontinuity stretching around the hemisphere, also later abandoned, betrays a similar hasty conceptualization, no doubt informed both by the vision of an uninterrupted circumpolar communication system and by the desire to keep the concept theoretically simple. In short, the original description of the polar front reinforces the assertion that the Bergen meteorologists were consciously thinking of the requirements for a weather service, which in turn related to mobilizing the international discipline to accept and use their meteorology.

Marketing the Polar Front:
The Micropolitics of Conceptual Change

In constituting the polar front, the Bergen meteorologists recognized an opportunity to incorporate their earlier innovative methods and models in a comprehensive framework for theory and practice. They believed that the polar front provided them with a basis for erecting a whole new system of meteorology. Solberg's claim is typical of the group's youthful enthusiasm: "We have, in fact, plans under consideration to write a completely new meteorology, a *Meteorologie der reinen Vernunft,* indeed why not also . . . *der praktischen,* to corrupt and plagiarize Kant. Our latest finding, which will surely be of great consequence, invites this."[45] Indeed, they quickly adopted for their efforts the catchall phrase "polar front meteorology"; it emphasized that the Bergen school was offering a new approach. When they began writing up their findings, they turned not to Kant, but to Darwin, for a parallel on which to base a title: "On the Origin of Rain."[46]

Bjerknes continued, as before, to spread the new methods and ideas being developed in Bergen, only now he incorporated these within the polar front meteorology and his advocacy of a circumpolar weather service. After lecturing in Norway he began thinking of "packing his suitcase to go on the road to advocate an international circumpolar service."[47] He sent Rossby home to try out the package on the Swedish meteorologists. Rossby pointed out the polar front's "great importance" for understanding the formation of cyclones and therefore for the prediction of weather. He then discussed the pro-

45. Solberg to Hesselberg, 28 March 1920, NMI.

46. References to "On the Origin of Rain" in Bergeron to Solberg, 11 May 1920, HSP, and Bergeron to V. Bjerknes, 10 January 1921, VBP. The book was never finished; instead the school's results were published as articles through the 1920s.

47. Solberg to Hesselberg, 28 March 1920, NMI; "Polar fronten," *Bergens Aftenblad,* 5 May 1920.

posal for a circumpolar weather service, focusing especially on Bjerknes's idea of establishing an international central bureau for preparing forecasts of the weather several days in advance by using the polar front.[48]

Bjerknes next began organizing two conferences, to be held in Bergen over the summer of 1920, at which he hoped to reveal to meteorologists of other lands both the Bergen meteorology and the plan for a circumpolar weather service. To accommodate both the official international scientific-cultural boycott of the Central Powers and his own desire to maintain links between the two opposing groups, he first invited scientists from Britain, France, Iceland, Norway, Sweden, and the United States for one meeting at the end of July; then those from Austria, Finland, and Germany were to join the Norwegian and Swedish meteorologists in early August.[49]

Nations not represented at the conferences received offers from Bergen meteorologists for personal instruction and information. To the Soviet envoy in Sweden they wrote letters describing the circumpolar weather service and offering assistance for reorganizing the Soviet weather service: "The proposal requires for its realization assistance from all countries of the temperate zone of the Northern Hemisphere. The moment might appear less suited for international enterprises of greater dimensions, but the rapid progress of air traffic during the war had made a corresponding development of the weather services a leading question, the solution of which depends upon international collaboration."[50] Claiming that their new findings, when used in a circumpolar weather service, could provide forecasts of a sort hitherto unthinkable, they tempted major Northern Hemisphere nations to adopt their plan. Jacob Bjerknes wrote to the Japanese Minister in Sweden and Norway in the following enticing manner: "Our recent results seem to promise great progress in weather-forecasting. Besides making the forecast for the next day much more correct and detailed, they also make very probable that forecasts for several days or perhaps weeks might be attained, provided that corresponding international arrangements could be made."[51]

Encouraged by the summer conferences Bjerknes intensified his efforts. Asserting that they should "strike while the iron is hot," he asked Solberg to assist him in elaborating the mathematical basis for the new meteorology and in interesting others in their work. This

48. Summary of Rossby's talk in *Geografiske annaler* 2 (1920), 184.

49. NMLÅ *1920–1921*, pp. 20–21, 38.

50. Veirvarslingen paa Vestlandet to Lemonosov, 28 August 1920; see also Rossby to Lemonosov, 27 December 1920; copies, VpVA.

51. J. Bjerknes to Sir Hicki, 2 April 1920, copy, VpVA.

suited Solberg, who wanted to withdraw from the arduous forecasting work and resume studying mathematical sciences; he could continue in the Bergen school as theoretician and propagandist. Bjerknes wrote to Solberg that the conference was "very encouraging and if we can follow up our victory immediately, then perhaps in the course of a couple of years we will organize the polar front meteorology."[52] He envisioned sending Jacob to Belgium, England, France, Germany, and Holland in the fall and going himself, with Solberg, to Washington after Christmas to "teach" the new meteorology to Marvin, the head of the U.S. Weather Bureau. Bjerknes referred in fact to his assistants as "apostles" who must spread the new meteorology around the world.[53]

The Bergen school's younger members shared its founder's zeal. They dubbed themselves "apostles," "missionaries," "comrades-in-arms," and "warriors" when discussing their efforts to convert foreign meteorologists. Naturally, much of this language was merely enthusiastic playfulness, but it also reveals the determination of the young to convince older, established colleagues in many countries to adopt the Bergen results. Like Bjerknes, they understood that for their work to have an international impact, they would have to teach and persuade. Bergeron was the most enthusiastic and perhaps the most vivid in his language: "The time has come for the army north of the p[olar]-front to concentrate its forces [*stridskrafter*] and, following Napoleon's celebrated example, with superior force strike at the enemy on small sectors of the front one at a time."[54] Solberg—"our highly valued French 'sapper' "[55]—reported from Paris a plan for diffusing the polar front meteorology, one that "will leave all of France permeated [*gjennemsyret*] with the Norwegian poison."[56]

Jacob, who tended to be more passive, for he believed that the truth would win out in the end, regarded himself nonetheless as an "apostle" when he attended conferences and sojourned in other meteorological centers.[57] In his own quieter way he assessed potential supporters: of several foreign meteorologists he wrote that one "is first and foremost kind, but that is not sufficient to be a polar front meteorologist"; of another who had worked for twenty years under

52. V. Bjerknes to Solberg, 2 August 1920, HSP.

53. Referred to by J. Bjerknes to Solberg, 25 June 1920, HSP. Vilhelm had used the expression "missionary among the heathen" earlier to describe his campaign for cgs units.

54. Bergeron to Ernst Calwagen, 4 October 1921, TBP.

55. Ibid., 20 September 1921, TBP.

56. Solberg to V. Bjerknes, 14 May 1921, VBP.

57. J. Bjerknes to Solberg, 12 August 1922, HSP. Bergeron (interviews) claims that Jacob did not consider special efforts necessary for spreading the Bergen meteorology.

the elderly Berlin meteorologist Gustav Hellmann, "Can a Hellmann meteorologist become a polar front meteorologist?" and of a sharp but at times difficult colleague, he noted that other foreign meteorologists might be nicer, but these less intelligent persons were not the kind of allies they needed.[58]

Marketing new ideas, techniques, and research programs is of course common in science, but there were particular reasons for it in the case of the young Bergen school. Skepticism and opposition greeted them more often than not. The reasons are many—too many to treat here—but part of the problem was that they were trying to market an incomplete product. Theoretical underpinnings for the new models had yet to be worked out; heavy work loads left little time for writing and publishing. To adopt a substantially different form of analysis and thought, meteorologists needed to study the results carefully, which without detailed publications, they could not do. Bergen meteorologists' preliminary publications and lectures at conferences, moreover, tended to be dogmatic. Also, adopting the Bergen results meant adopting a new form of analysis; experience showed that much of the craft involved could be learned only through direct contact with somebody who had mastered the methods. All this was in addition to normal institutional inertia and reluctance to learn new techniques. The change amounted, after all, to a total shift in how cyclones were conceived. The group felt at times that the resistance also had a prejudicial component: How could anything important come from a little country like Norway—and from such young, largely unknown scientists?[59]

Bergeron remarked in relation to one of his "battle plans": "even the linguistically ignorant [*språkokunniga*], conservative, imperialistic sons of John Bull must be capable of understanding the value of

58. J. Bjerknes to Bergeron and Calwagen, 13 February 1923, TBP; J. Bjerknes to Bergeron, 14 November 1923, TBP.

59. Francis W. Reichelderfer says this attitude was prevalent both in America and Germany during the 1920s. After a sojourn in Bergen in the early 1930s, he visited Germany, where many meteorologists admitted they had read the classic Bergen school articles when they first appeared but did not believe anything really important could come from Norway. He had met the same reaction earlier in America. He himself had almost immediately recognized the significance of the early Bergen papers because of his own experiences as a U.S. Navy airforce meteorologist. First, when accompanying Billy Mitchell on one of the demonstration bombings of obsolete ships, they narrowly averted disaster during the passage of a strong cold front. Then, as meteorologist for the navy's airship program in the mid-1920s, he appreciated again the significance of focusing on surfaces of discontinuity, which could, and did, wreak havoc on airships. Reichelderfer introduced the Bergen school methods into the navy's weather service and then, in the late 1930s, also into the U.S. Weather Bureau, when he was named director. (Interviews, F. W. Reichelderfer, Summer 1978, Washington, D.C.)

observations and methods other than those they developed during the war, or knowledge of the *existence* of countries and civilized nations north of latitude north 60°."[60] Austrian meteorologists accused the Bergen school of not giving sufficient credit to their contributions, which they claimed were precursors or even independent prior discoveries. Some of these criticisms tended toward accusations, and not without some emotional heat. The leaders of Austrian meteorology, Felix Exner and Heinrich von Ficker, were especially troublesome.

On one occasion while visiting Germany, Exner became "completely enraged [*geradezu erragister*]" by the Bjerknes methods; he vented his feelings in rather ugly expressions.[61] Undoubtedly the total defeat and dissolution of the Empire, the economic catastrophe that all but shut down scientific work, the international boycott, and the recognition that for the moment Vienna had ceased being a leading center for meteorology contributed to his intemperate private and public outbursts. His colleague, von Ficker, explained the opposition to the Bergen school: "The Norwegian school . . . has broken the Austrian school's hegemony in meteorology."[62] Be that as it may, Bjerknes and the others understood that Exner, as editor of *Meteorologische Zeitschrift*, wielded sufficient power to be a dangerous opponent. Sometimes personal intervention, discussion, and concession helped, as in Jacob's and Vilhelm's soothing of von Ficker, who accepted much of the early Bergen school's results but was "angered on behalf of the Austrians over the lack of citations. It's a question of making up with him, now when he is becoming the big boss [*storpamp*] in Berlin after Hellmann."[63]

In summing up the opposition of the early years, Bjerknes remarked that most meteorologists "have much trouble in thinking in three-dimensions" and that while "many lack understanding of the Bergen results others lack the desire to understand."[64] Differences of opinion naturally arose among the Bergen meteorologists on how to present their results. On occasion, they even disagreed on what results actually had been won. Immediately after the polar front was established, there were many uncertainties as to the character and varieties of the fronts, the specifics of cyclonic development and evolution, and the reality of several secondary discoveries. Criteria for

60. Bergeron to Calwagen, 20 September 1921, TBP.
61. Bergeron to H. Bjerknes, 21 January 1923, VBP, quoting Ludwig Weichmann, who was sympathetic to the Bergen school.
62. V. Bjerknes quoting von Ficker in letter to C. W. Oseen, 16 July 1929, CWO.
63. J. Bjerknes to Bergeron, 14 March 1923, TBP.
64. V. Bjerknes to Bergeron, 21 May 1924, copy, VBP.

spreading the results played a role in the processing of "discoveries"—that is, the transforming of members' insights, postulates, and empirical claims into a reality they all could fully accept.

The establishing of the reality of secondary cold fronts, which were observed in 1919 and 1920, is one example. These parallel fronts—one in back of the other—within the polar air mass implied a relativity of the polar front; that is, they contradicted the assertion that the polar and tropical air are absolute qualities and that therefore in the northern hemisphere a single polar front separates them. In addition, when foreign meteorologists were having sufficient difficulty in accepting the basic idealized 1919 cyclone model based on fronts, such details appeared to be unnecessary complications. Asked by foreign meteorologists about the possibility of such secondary cold fronts, Jacob categorically denied their existence. Bergeron went around at the same time delineating such structures on weather maps. Even when some members (Ernst Calwagen and Anfinn Refsdal) published maps showing secondary old fronts, Jacob continued to deny their existence. According to Bergeron, Vilhelm and Jacob "prohibited 'the rest of us' to pretend to the outer world that something like that [secondary cold fronts] exists, because it might prejudice the hegemony of the one and unitary, grace-saving polar front."[65]

Vilhelm continued claiming a single polar front well into 1921, while the others tried to claim that breaks occur in the front. Bergeron and Solberg discussed the possibilities of relative polar fronts, which would mean that in addition to breaks within the main polar front, transitional polar fronts may occur separating, say, artic from polar, or tropical from subtropical air. Many new terms—*maritime front, Moroccan front*—appeared and disappeared in their private correspondence. Bergeron also began developing a system of upper fronts and—much to Jacob's chagrin—modifying the idealized warm-sector cyclone. Most of these developments were not officially acknowledged at first. When revising for publication the first full-length post–polar front article, Solberg removed many discoveries and details. He decided that the best strategy for convincing foreign meteorologists entailed keeping everything as simple and elegant as possible: "the crystal-clear drops [of water] seem more refreshing to a thirsty soul than a whole flood of muddy water."[66]

But demands for further improvement in accuracy and specificity of the forecasts eventually countered the urge for simplicity and established the reality for the Bergen school of some of these concepts.

65. Bergeron to Solberg, 20 June 1924, HSP.
66. Solberg to V. Bjerknes, 20 June 1921, VBP.

This drive toward further sharpening of the methods arose from new circumstances at home. Indeed, even as they thought of international conquests, they found that the entire enterprise was in jeopardy. Support from the Norwegian government was not at all reliable, for by late 1920 a major economic depression had developed in Norway. To keep their institutional basis intact and to justify their requests for support, the Bergen meteorologists had to integrate weather into the economic sphere yet further, especially into fishery. And as part of this process, they modified their initial atmospheric models and devised new ones.

PART V

RATIONALIZING
THE WEATHER
(1920–1925)

[10]

Fishery, Occlusion, and the Life Cycle of Cyclones

NORWEGIAN fishermen always needed protection from hazardous conditions, for the west and north coasts of Norway are among the most stormy in the world. Attempts in 1909 to aid fishermen by supplementing their age-old folklore with a storm-warning system had only marginal success. Norwegian fishermen began to desire and require a comprehensive daily weather forecasting service after 1919; at the same time, Bjerknes and his colleagues began to consider the forecasting needs of fishery as a major aspect of their enterprise. The economic crisis in 1920 prompted closer relations.

By late 1920 the economic boom, which had begun with runaway wartime inflation, finally ended. The Norwegian economy entered a severe depression that continued to afflict most Norwegians for much of the interwar period.[1] Because of the extraordinary role of her merchant fleet during the war, Norway had become by 1919 a creditor nation for the first time in her modern history. Once restrictions on imports were lifted, a massive wave of buying drained this credit. By the spring of 1920 the impact of the shattered world trade system began to be felt. In the summer, after the collapse of the international freight market, Norway's economy seemed to fall apart. As the British trade minister in Christiania reported to London, "It was evident at the close of 1920 that a difficult period was impending, but it is

1. A general account of the entire period can be found in Hans Fredrik Dahl, *Norge mellom krigene* (Oslo, 1971); specificly for this chapter, I am also indebted to Berge Furre, *Norsk historie, 1905–40* (Oslo, 1972), esp. pp. 124–216ff; Fritz Hodne, *An Economic History of Norway, 1815–1970* (Trondheim, 1975), chaps. 14–15.

doubtful if many persons then foresaw the extreme nature of the coming depression. . . . From every point of view the past year has proved a time of crisis: it has left its mark upon all sections of society. . . . The fisheries, shipping, industry and commerce suffer from an almost unexampled depression."[2] This depression, though crucial in forging meteorology's new relationship to fishery, was not the only determinant. Another was the major structural change that had occurred during the war in this important segment of the Norwegian economy: the composition of the Norwegian fishing fleet had been transformed, and the introduction of so many privately-owned motorized boats contributed to the postwar crisis.

Motor-driven fishing boats began gradually joining the Norwegian fleet in the early part of the century. The new boats, large and covered or small and open, but all with diesel motors,[3] allowed the fishermen to venture further out to sea and follow the catch along the coast. War brought soaring fish prices. Germans and Britons competed for Norway's catch, the latter hoping thus to prevent the Germans from obtaining much needed food imports. Widespread anticipation of British buying drove up Norwegian fish prices in 1915 to six times the 1914 levels. The British continued buying fish through 1917, after which the state began purchasing the fish because no other market existed. Although the state tried to keep the minimum guaranteed prices as low as possible, it continued buying much of the surplus through the 1918–19 winter.[4] Prompted by the high prices for their catch, many Norwegian fishermen had purchased small motorized boats. Whereas in 1910, for example, there were 505 open motor vessels, 2407 decked motorized vessels, and 3779 sailing vessels registered, by 1920 the respective number of boats registered was 6066, 8799, and 327.[5] The effect on the industry of all these small, motorized open fishing vessels purchased on credit during the wartime inflationary boom was profound.

The postwar contraction of the market for Norwegian fish began when the Central Powers, already owing huge sums from wartime

2. Great Britain, Dept. of Overseas Trade, *General Report on the Industrial and Economic Situation of Norway in December, 1921* (London, 1922), p. 6.

3. Hodne, *Economic History of Norway*, pp. 86–88; Eivind Thorsvik, "Mekanisering av fiskerflåten," in Norwegian Museum of Technology, *Volund 1972: Beretning for 1971*, pp. 9–131.

4. Wilhelm Keilhau, *Norge og verdenskrigen*, Carnegie Endowment for International Peace: Economic and Social History of the World War, Scandinavian Series (Oslo, 1927), pp. 118–37; Per Vogt, *Jerntid og jobbetid: En skildring av Norge under verdenskrigen* (Oslo, 1938), pp. 61, 168–69.

5. Furre, *Norsk historie*, pp. 22, 144; *NOS XII 245*, Table 116, "Registered Fishing Vessels."

fish purchases, could not pay cash because of their own economic turmoil. After the revolution, the important Russian market was closed. The Norwegian government was increasingly pressed to find funds to buy the surplus fish stocks from the fishermen.[6] Further contraction occurred when Spain and Portugal retaliated against Norway's prohibition of importation of alcoholic spirits. Cod, the backbone of the industry, fell in price from 160 øre/fish in 1919 to 55 øre/fish in 1921; equally catastrophic drops hit prices of other major fish.[7] Meanwhile the cost of gear, repairs, equipment and fuel continued to climb. The cost of operating the mechanized fleet virtually canceled any possible gain:

> One of the most difficult problems with which Norway is faced arises out of the position of the fishing industry. The fishermen . . . are in a condition of great distress owing to the heavy cost of supplies—(coal, petroleum and all kinds of gear), and, particularly, owing to the failure of their usual markets.[8]

> The difficulties under which the fishing industry laboured in 1920 have continued during the present year, and the distress in the fishing districts has remained acute. . . . The heavy fall in prices obtainable for fish products has not been counterbalanced by a corresponding reduction of the cost of gear, petroleum, etc., so that the fishermen have in general been unable to work at a profit, or even to cover expenses.[9]

Recall that about one-third of Norway's population then lived on the coast and derived some part of its livelihood from the sea. The labor unrest and increasing radicalization of the Norwegian work force that had begun toward the end of the war spread among the fishermen.

To prevent financial ruin, fishermen believed they had to increase their catch to compensate for the decline in value.[10] Loan payments on motorized boats and new gear purchased during an inflationary period still had to be made.[11] Operations had to be as efficient as possible. Although the fishermen who ventured into the North Sea always needed storm warnings, which could save their lives, they now required precision daily forecasting to squeeze the most from their

6. Great Britain, Dept. of Overseas Trade, *General Report on the Industrial and Economic Situation of Norway, December, 1920* (London, 1921), pp. 15–16.

7. Great Britain, Dept. of Overseas Trade, *General Report, December, 1921*, pp. 18, 21–22. At this time, cod made up 60 percent of the catch's value; the decline in herring prices was even more precipitous, *NOS XII 245*, Diagram 29, "Landed Value of Fish."

8. Great Britain, Dept. of Overseas Trade, *General Report, December, 1920*, pp. 15–16.

9. Ibid., *December, 1921*, p. 18.

10. Furre, *Norsk historie*, p. 144.

11. Hodne, *Economic History of Norway*, p. 412.

precarious business. With fuel so costly, they could ill afford to travel out to sea when conditions would be unfavorable for fishing or to be driven home by an unexpected storm.[12] Likewise, unnecessarily to remain in port or refrain from venturing to the best fishing areas during the peak seasons would also have to be avoided. Loss of gear or damage to fishing vessels in severe weather was equally troubling with replacement costs so exorbitantly high relative to income from the catch.

After having guaranteed the purchase of the catch during the last years of the war and in 1919, the government tried to diminish its expenditures in this sector. But having moved significantly away from a laissez-faire position and having to react to the increasing radicalization of the Norwegian work force, the state was forced to involve itself in trying to alleviate the worst effects of the growing economic crisis. In 1920 and 1921 the government implemented emergency measures to aid fishermen, including appropriations of 5 million crowns to reduce the cost of petroleum for the vessels and 2.3 million crowns to help poor fishermen purchase gear. The long-awaited state-guaranteed Fisheries Bank opened and began providing loans for purchase, reconstruction, and repair of fishing crafts and for other needs of individual fisherman. Government also widened the accident insurance benefits for fishermen and began providing pensions for fishermen and their families in case of incapacity or death. Studies were also initiated to reform and to coordinate insurance for fishing crafts, which continued to soar in value.[13] Effective weather forecasting could save the state considerable expense.

The Bergen Meteorologists Discover Fishery's Significance

After 1920 the Bergen meteorologists suddenly turned to fishery as their most important social concern. In his September 1919 report on the Norwegian weather forecasting services, Bjerknes said very little about predictions for the fisheries. He expressed hope that the daily

12. "Veirvarslingen i 1921," *Morgenavisen,* 4 January 1922; "Aktuelle problemer i veirvarslingstjenesten i Bergen," ibid., 23 August 1923.

13. Great Britain, Dept. of Overseas Trade, *General Report, December, 1921,* pp. 19–22; *Statistisk Årbok for Kongeriket Norge, 1924* (Oslo, 1925), p. 79, Table 69: the total value of the fleet increased from 40,886,075 crowns in 1914 to 138,063,907 crowns in 1919 and remained near that level in the early 1920s. Bjerknes and his colleagues were well aware of the inflated prices of fishing vessels and gear from direct communications with the General director of fisheries; see Fiskeridirektøren to Det geofysiske institut, 15 March 1920, VBP. The value of the nation's fishing gear, according to the director, had increased from about twenty million crowns in 1914 to about fifty million kroner in early 1920.

summer forecasting service could be extended year round. He cited the coming needs of aviation as the primary justification for such an expansion and pointed out that the expanded service could at the same time encompass and improve the prewar coastal storm-warning service for fishermen. When the actual summer forecasting season ended in September 1919, forecasting was limited to issuing warnings to various points along the coast, whenever winds above strong gale force were expected. While discussing further cooperation between the fisheries and the weather forecasting service, Jacob Bjerknes informed the board of directors of the fisheries (*Fiskeristyret*) that the only forecasts available during the winter were to be those of strong gales or storm-force winds that could cause damage and loss of life. And these, he thought, would be sufficient to meet the fishermen's needs. Should the fishermen desire daily forecasts, these could in principle be prepared, but no funds had been appropriated or were available to pay for the extra telegrams to the fishing stations.[14] The Bergen meteorologists did not yet believe expanded services for fishery were essential.

Some fishermen believed daily forecasts during the winter could be useful.[15] Having become used to receiving these daily forecasts and to depending on them, some were prepared to pay for the cost of receiving daily telegrams to their fishing stations. The fishermen at Sørgjæslingerne decided to pay for the telegrams themselves for that winter, as did the fishermen who set out from the Egersund fishing station during the spring herring season.[16] Others were less sure that scientists sitting in Bergen could offer them forecasts better than those they could arrive at with their own folklore. By midwinter, however, the Bergen meteorologists had won most fishermen's confidence.

By midwinter the Bergen meteorologists also learned that they very much needed the support of the fishermen. Having worked with leaders of military and commercial aviation through the summer and fall of 1919, Bjerknes and Hesselberg wrote budget proposals justifying increases in personnel, salary, and operating expenses with aviation's projected needs. They based virtually all planning for expanding the weather services in southern Norway on aviation's needs because they assumed Norway would start investing large sums in

14. J. Bjerknes to Fiskeristyret, 16 October 1919, copy, VpVA; V. Bjerknes, "Betænkning om den norske veirvarsling og dens utvikling," KUD/D, folder: "Værvarslinga på Vestlandet," file: "Budsjettsaker 1916–1946."

15. J. A. Hemmingsen (representing fishermen from Sørgjæslingerne) to J. Bjerknes, 4 October 1919, VpVA.

16. NMIÅ *1919–1920*, p. 35.

aviation and aviation-related services. But this was not to be. In late
December 1919, Hesselberg learned informally that the Ministry of
Church and Education was going to recommend trimming the budget
that they had proposed.[17] Upon learning this information Bjerknes
became angry. He responded rashly. Without an increase in personnel, he claimed, the few forecasters would not be able to devote any of
their time to research in support of the practical work. In that case he
would prefer to close down the entire Bergen forecasting service; he
and his assistants could devote themselves to the study of the enormous amount of data being assembled from the wartime European
field weather services.[18] Hesselberg turned to the leaders of aviation
for assistance in mobilizing support against the cuts,[19] but aviation
interests in Norway proved to have insufficient political strength to
guarantee the short-term growth and success of Norwegian meteorology. What saved the proposed budget was an unfortunate
accident.

On the night of 14 January 1920 a severe storm in the Titran area
of the Trøndelag coast that was not predicted killed twenty-one fishermen and resulted in loss of gear alone valued at half a million
crowns.[20] The tragedy was heightened by the memory of an even
greater accident some twenty years earlier in the same area in which
over one hundred fishermen lost their lives in one night. Primarily
smaller, privately owned boats were out in the fishing grounds—
those of fishermen in greatest economic need. The larger commercial
boats had not ventured out because of the poor market prices and
because of strikes. The incident underscored the plight of the fishermen and prompted extra legislative actions to organize diverse
emergency measures. The storm itself, as Bjerknes pointed out, was
of a type that comes very quickly from the Arctic Sea, whence no
weather observations were available.

When the Storting considered the matter, representatives from
various industries, most notably the fisheries, came to the support of
the forecasting services; the Storting approved most of the original
budget and granted an emergency extra appropriation to help the
fisheries immediately. The government declared it "extremely important" that the forecasting service be organized to coincide with the
major fishing seasons.[21] Bjerknes was elated: "But the main thing is

17. V. Bjerknes to T. Hesselberg, 28 December 1919, NMI.
18. Telegram, V. Bjerknes to Hesselberg, 23 December 1919, NMI.
19. Hesselberg to Oberst Gustaf Grönner, 3 January 1920, copies, NMI.
20. "Forespørsel fra rep. Sivertsen angaande regjernings foranstaltninger i anledning av stormulykken paa Titranhavet," *Forhandlinger i Stortinget 1920* (Proceedings of
the Norwegian parliament), 7ende del. A.
21. "Om bev. til meteor. institutioner," *Forhandlinger i Stortinget 1920–1921*, indst. s.
no. 35.

the general approval, especially from the representatives for the coastal population and the fishing interests, [which] from now on assures our undertaking's future."[22] Later, after discussions with the Fishery Directorate, an additional thirty thousand crowns was approved as part of the emergency measures for the fisheries, which would cover the cost for sending forecasts daily by telegram to 360 fishing stations along the coast.[23] Moreover, funds were rapidly provided for setting up a weather station equipped with a wireless telegraph on the remote polar island of Jan Mayen, located in the blind spot along the path of these severe winter storms.

Bjerknes soon began visiting fishing stations to learn more about how the forecasting could be made more effective. He became "a regular fixture [*et fast inventar*]" at meetings of fishermen's organizations: lecturing on the science of forecasting and listening to the problems facing the fishermen.[24] The fishermen, Bjerknes heard, needed forecasts that could be used in organizing their work, forecasts detailed and precise enough to help them decide when to go out and where to go.[25] During the important winter fishing season, storms are apt to follow one after the other; the job for the forecaster was to predict where these cyclones might affect the coast and where and when it might be safe to fish. To be able to use the periods of safe weather emerged as a goal as important as avoiding dangerous weather. The forecasters worked to increase the number of warning signal masts and explored ways of introducing wireless communications to the fishing stations and to boats at sea.

Bergen meteorologists worked with the Fishery Directorate to compile lists of the comparative vulnerability to wind and rough sea of the different types of boats and gear used in various fishing districts. They needed to determine up to what threshold wind speeds and directions fishermen could work in relative safety.[26] The school tried as well to obtain "the most exact and complete observations possible" and to give them "the most painstaking analysis possible" in order to understand and predict the physical changes occurring in the atmosphere.[27] Complicating the problems facing the forecasters was the

22. V. Bjerknes to Axel Wallén, 27 February 1920, copy, VBP.

23. NMLÅ *1920–1921*, pp. 16–17; Hesselberg to Fiskeridirektøren, 5 May 1920, copy, NMI.

24. Bjerknes to H. U. Sverdrup, 15 October 1921, copy, VBP; "Aktuelle problemer i veirvarslingstjenesten," *Morgenavisen*, 23 August 1923.

25. V. Bjerknes, "Om veirforutsigelse som fysisk problem," *Naturen* 47 (1923), 45–74; "Veirvarslingen i 1921," *Morgenavisen*, 4 January 1922.

26. Værvarslingen på Vestlandet (Finn Spinnanger) to Fiskeridirektøren, 16 December 1922, including the questionnaires sent to various fishing districts, and 9 August 1923, copies, TBP.

27. "Prof. Bjerknes' svar paa direktör Sandströms artikkel," *Bergens Tidende*, 29 December 1923.

continued need to rely on telegraphy to spread forecasts to the many fishing stations. A storm expected in the afternoon would have to be predicted by 9 o'clock the previous evening, based on that afternoon's map analysis. Forecasts based on the next morning's observations could not reach the fishermen in time.[28] To fulfill their pledges to help the fishermen—and the state—the Bergen meteorologists would have to find better means of comprehending and predicting cyclonic motions and evolution. If the fishermen were disappointed in the forecasts, the state would, in this time of economic crisis, substantially cut the meteorological budget.

Seclusion, Occlusion, and Evolution

Efforts to improve the storm warnings and weather forecasts prompted further conceptual change and established the "reality" of still-questioned models and concepts. Forecasting's priority over other concerns caused the occlusion process as well as secondary cold fronts and nonunitary polar fronts to become established as fundamental principles in the Bergen meteorology. Finally, by 1922 the group arrived at a preliminary schema for the evolution of cyclones.

Recall that during the fall 1919 forecasting season the Bergen meteorologists increased the number of observations periods and map analyses to three times daily, which in turn afforded them opportunities to study more closely the behavior of the new cyclone model. They then saw that secondary cyclones can form on the cold front of an existing cyclone. At this time Bergeron began to notice some additional curious events. On his weather maps, he noticed that between 23 and 30 September a cold front appeared to overtake a warm front, thus making the warm sector between them smaller and smaller. He saw the same thing again on 18 November. This time, moreover, the entire warm sector seemed to disappear! Was this a phenomenon actually occurring in nature, or was it just a construct arising from the map drawer's analysis or imagination? Bergeron remained in the forecasting office very late that evening, contemplating this enigma. Finally, he says, he sketched a hypothetical explanation at the bottom of the map; it showed schematically the cold front lifting the entire warm-air sector above the ground.[29]

Neither Bergeron nor his colleagues considered this finding significant at the time: they apparently saw no deeper implications and

28. Værvarslingen på Vestlandet to Fiskeridirektøren, 16 December 1922 and 9 August 1923, copies, TBP.

29. Interview, Tor Bergeron, 9 June 1976; these details are repeated in the numerous historical recollections of the Bergen school.

could not define the precise nature of this phenomenon, if indeed it actually existed. Bergeron's own letters from the late fall and winter, which contain detailed accounts of the work in Bergen and especially of his own many activities and ideas, offer not a single indication that this discovery had been made. In one letter, to his Swedish colleague and friend Oscar Edlund, Bergeron discussed the meteorological details of the storm on 18 November but said nothing about this phenomenon.[30] Vilhelm Bjerknes's letters from this period also contain no references to this phenomenon although he reported all the achievements of the fall forecasting season. In a letter recommending Bergeron for a position in Sweden as a weather forecaster, Bjerknes praised his great ability both to analyze weather maps and to link map analysis with observation of the sky; Bergeron's work on visibility also received praise.[31] Nothing was said of his hypothesis.

It may well be that upon discussing this finding with his Bergen colleagues, Bergeron met only skepticism. He had no means of reproducing his insight; any given analysis of the weather using the new and still-developing methods was necessarily a fluid, tentative interpretation. What would constitute proof that the warm sector disappeared? And if this disappearance indeed had occurred, how could they know whether it represented any fundamental atmospheric process? To what question or problem was Bergeron's "discovery" or hypothesis an answer? Jacob Bjerknes, for example, resisted any modification of his original cyclone model composed of warm and cold fronts enclosing a warm sector.[32] He preferred to conceive this configuration as immutable; he tended to find warm-sector cyclones with the same sector angle separating the two fronts in all the observations. Yet by the summer of 1920, Bergeron resumed active inquiry into this phenomenon and even convinced Jacob that this coming together—"*sammenklapping*"—of the two fronts in a cyclone must be studied further. Two related developments helped make Bergeron's hypothesis a viable problem.

From their analyses of daily weather maps and from their work on the polar front, Bergen meteorologists gradually came to accept in 1920 a process, called "seclusion," in which the cold front on the map overtook the warm front at a point well south of the center of the

30. Bergeron to Oscar Edlund, 26 November 1919, TBP.

31. V. Bjerknes to Wallén, 20 December 1919, copy, VBP.

32. Interview, Bergeron; in a letter to H. Solberg, 20 November 1920, Jacob still tries to stress the original ideal warm-sector cyclone of 1919 in a map analysis, HSP; Bergeron related in an informal unpublished recollection, "Minder fra Bergenmetodernes utvilking etter 1918 og de den gang medagerende," that Jacob did not like to see his model "deformed" but ultimately went along with the change. I am indebted to the late professor Bergeron for a copy of this memo.

cyclone; the supply of warm air flowing into the warm sector is pinched off, leaving a pool of warm air encircled by cold air.[33] Seclusion seemed rather to be the consequence of a mechanical process acting on the cyclone than to be a part of some inherent, natural life-cycle process. Mountains, for example, can slow down or stop the forward movement of a warm front and thereby permit the cold front to catch up and to intersect the former discontinuity at some point below the cyclone's apex. Cyclones trying to pass over southern Norway frequently exhibited this process, in which the warm sector would "shrivel up [*skrumpe inn*]." In such situations they observed that secondary cyclones occasionally formed at the point of intersection. Seclusion appeared, therefore, to be connected with the birth of new cyclones; it also seemed to imply the death of old cyclones by the cutting off of the supply of warm air. Still, the process was only vaguely understood.[34]

Bergeron realized that a relation existed between what he had seen in fall 1919 and seclusion. By collecting during summer 1920 all available data related to his original November 1919 case and by analyzing them as minutely as possible, he began to develop the notion of "*sammenklapping*."[35] He noted that after the initial disappearance of the warm sector, as the two fronts came together, the rain area originally connected with the leading warm front began both to shrink and to move well in advance of the remaining line of discontinuity at the surface. And yet little if any rain could be associated with this remaining discontinuity. Rain was falling many miles in front of this line, where on the surface there was no indication of any discontinuity in any weather element. The band of rain seemed to be falling in the middle of a cold air mass. Bergeron inferred from this phenomenon that after the *sammenklapping*, an "upper front" existed that, having been lifted off the ground, was carried by the advancing cold air further away from the remaining surface front. This upper front had all the characteristics of a warm front except that miles behind this upper front's base, at ground level, a discontinuity could be delin-

33. V. Bjerknes, "Om vær- og stormvarslinger og veien til at forbedre dem," *Teknisk Ukeblad* 38 (1920), p. 305; idem, "The Meteorology of the Temperate Zone and the General Atmospheric Circulation," *Nature* 105 (24 June 1920), 524.

34. V. Bjerknes, "Om den almindelige atmosfäriske cirkulation," lecture in Copenhagen, October 1920, VBP; J. Bjerknes, untitled lecture at a meeting of the Italian Meteorological Society, Venice, 1920, JBP.

35. Bergeron to Solberg, 21 August 1920, HSP. The map of 18 November 1919 showing an upper front that was used in several historical articles by Bergeron and others is actually the reanalyzed map, not the original from the date; Bergeron interview, 9 June 1976; Bergeron, in an untitled lecture from 17 September 1953, confirms that the drawing he often presented was the reanalyzed map.

eated with little rain associated with it.[36] Although Bergeron convinced himself of the existence of upper fronts, evidence indicates this notion was not immediately adopted either in forecasting work or in presentations of the Bergen meteorology.

By the fall and winter forecasting season of 1920–21, the Bergen group fully appreciated the need to predict storms and gales along the Norwegian coast more accurately. Although Bergeron was then forecasting in Stockholm as a "Bergen meteorologist," he remained in close contact with his Bergen colleagues, exchanging ideas and data, especially with his Swedish friend Ernst Calwagen, who assumed a major responsibility for leading the forecasting in Bergen from January 1921. All concerned agreed on the necessity of yet sharper analyses of the incoming observations and further refinements in the predictive methods. When a weak and not very threatening cyclone, with its warm and cold fronts, appeared on the weather map—for example, north of Ireland—how could the forecasters in Bergen predict whether this system would intensify into a major storm, die out, or reach them virtually unchanged? All these possibilities had been observed; some storms rapidly intensified, others did not. Fishermen and forecasters alike agreed on the necessity of predicting early and accurately the many cyclones with hurricane-force winds that reach the Norwegian coast during the fall and winter. To do so would require an understanding of a cyclone's evolutionary mechanism. Seclusion and *sammenklapping* seemed to hold part of the answer. As Bjerknes commented in spring 1921, "This process of seclusion and the phemomena accompanying it are exceedingly important from the point of view of the forecasts, and will therefore be made the subject of detailed treatment in later papers issuing from the Norwegian Weather Service."[37]

Calwagen instituted rigid systematization in the forecasting work. He strove to bring order in the chaos of details on the weather maps through constant refining of the analytic methods. He tried to identify—and predict—every minor squall sweeping in from the North Sea; he numbered each front and followed it from map to map. So he was able to help establish the "reality" of secondary cold fronts within the polar air: behind each front the wind was successively more northerly and colder. He and the others sought criteria for predicting intensification of individual fronts and cyclones. In addition to sharpening the painstaking "microscopic" analysis of the observations, they were

36. I am deeply indebted to Bergeron for providing me with this explanation of the work leading to the understanding of the occlusion process.

37. V. Bjerknes, "On the Dynamics of the Circular Vortex with Applications to the Atmospheric Vortex and Wave Motions," *GP* 2, no. 4 (1921), 80.

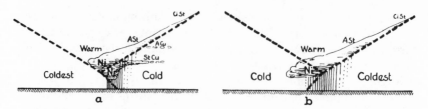

Fig. 22. Two models of the occlusion process (1922). The situation Bergeron first analyzed entailed less cold air advancing on a colder air mass (*b*). As the upper front continued to advance on the surface of the preexisting warm front, the area of rain also advanced so that, at the surface rain, was falling in the coldest air. From J. Bjerknes and H. Solberg, "Life Cycle of Cyclones and the Polar Front Theory of Atmospheric Circulation," *GP* 3, no. 1 (1922), 8.

aided in that the observations had become increasingly more detailed. In 1921 all observation stations began reporting considerably finer distinctions in types of weather and clouds. Methods of "indirect aerology" could be elaborated with greater sophistication; the "reading" of these observations as signifiers of processes occurring in three dimensions finally began to mature into a rigorous method of analysis. Bergeron paved the way in this development. Using indirect aerology as part of their more rigorous analyses, the Bergen meteorologists were able to delineate in 1921 several processes including the role of *sammenklapping* in the evolution of cyclones.

First, using the improved data and refined method of analysis, Bergeron was able to show convincingly in the forecasting work that two types of *sammenklapping* processes occur (Fig. 22), which one depending upon whether the cold air advancing from behind is colder or warmer than the slower moving cold air ahead of the original warm front. In the case of the November 1919 instance, the air arriving behind the cold front was warmer than the retreating cold air, so that when it collided, being warmer and therefore lighter, it moved up along the existing surface of discontinuity, forming an upper warm front. As this process of upper-front formation became clearer, the meteorologists began to use the word *occlusion* to describe the process.

While Bergeron concerned himself especially with the characteristics of the upper fronts, those in Bergen, in their the daily discussions of the current weather map, concentrated on the front that remains at the surface:

The problem of occlusion was very frequently discussed [spring and summer 1921]. . . . The cyclones which J[acob] B[jerknes] first studied seemed to be those of summer seasons, and warm and cold fronts were clearly distinguished. However, in fall and winter seasons many examples of occlusion were observed and some cyclones were without warm fronts at all. During my stay, they found strange fronts extending northwestward from the cyclone centers. In some occasions these strange fronts showed further recurvature from west to southwest. Many discussions were held and then it was concluded that this was the final stage of occlusion.[38]

This single line of discontinuity found at the surface, surrounded by cold air, and extending out from the cyclone center toward the north, proved more significant in the daily forecasting than drawing the upper fronts per se. This occluded front, and the occlusion process in general, provided the key for arriving at a model for cyclonic evolution, a model that embraced a number of aids for forecasting and that could also be the focus of theoretical inquiries.

Figure 23 illustrates the life cycle of cyclones that the Bergen school arrived at in 1922.[39] This schematic horizontal view shows the development and dissipation of a cyclone at ground level. Initially, two different air masses flowing in opposite directions—ideally a cold easterly current adjacent to a warm westerly one—are separated by a straight boundary, the polar front (Fig. 23a). A cyclone begins to form where a slight bulge appears toward the cold side, caused by a projecting warm "tongue." "The centre of the cyclone will be found at the top of the projecting tongue of warm air" (Fig. 23b). Warm air in this tongue begins ascending over the cold air, forming warm-front rain. Meanwhile cold air begins curving around the northern end of the warm tongue and in so doing lifting the warm air, causing showers, in advance of this cold front. "The newly-formed cyclone follows the current of the warm tongue eastwards, propagating like a wave on the boundary surface between warm and cold air." As the young cyclone propagates generally toward the east, the amplitude, in the horizontal north-south direction, of this warm wave increases (Fig. 23c). This stage of development corresponds to the earlier static 1919 Bergen cyclone model with its well-developed warm sector bounded by a steering surface (warm front) and a squall surface (cold front).

As the wave increases in amplitude, the cold air in the rear begins to pinch the warm tongue inward along "the southern outskirts" of the cyclone (Fig. 23d). When this cold air from the rear finally reaches the

38. Sakuhei Fujiwara, "The Men Who Had a Deep Insight into Nature: The Birth of the Norwegian School," (trans. from Japanese, mimeo., Vervarslinga på Vestlandet, 1971), pp. 6–7.

39. J. Bjerknes and H. Solberg, "Life Cycle of Cyclones and the Polar Front Theory of Atmospheric Circulation," *GP* 3, no. 1 (1922), 4–9.

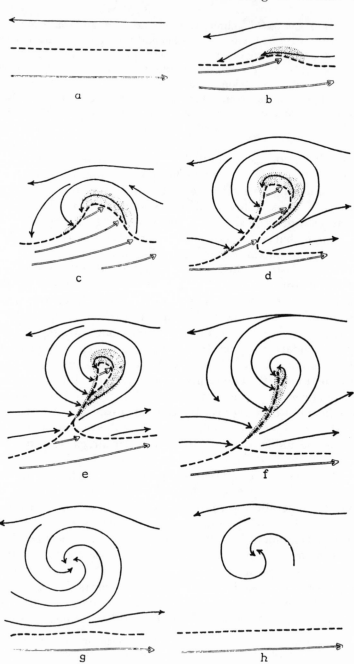

Fig. 23. The life cycle of cyclones (1922). From J. Bjerknes and
Solberg, "Life Cycle of Cyclones," p. 10.

cold air in front, the cyclone is cut off from its warm air supply and becomes "secluded" (Fig. 23e). Shortly thereafter the small remaining warm sector is also lifted off the ground, leaving at the surface only cold air. Because the cold air in the front and that in the rear of the cyclone will often have somewhat different properties, an occluded front remains (Fig. 23f). After some time the occluded front dissipates, leaving a cold vortex in which no lifting of warm air occurs; hence no major precipitation falls (Fig. 23g). Frictional forces finally dissipate the remaining vortex of cold air and the cyclone dies (Fig. 23h).

As long as warm air is ascending in a cyclone, energy is imparted to the system. The potential energy residing in a system in which cold and warm masses of air lie adjacent to each other becomes kinetic energy as the warm air is lifted. Hence, "all cyclones which are not yet occluded, have increasing kinetic energy." This increase in kinetic energy when a cyclone is found in a preoccluded state (Fig. 23a–d) is manifested by an increase in the intensity of the storm. After occlusion, some warm air continues to be lifted by the upper front, but essentially the reservoir of potential energy has been cut off. Without the coexistence of cold and warm air adjacent to each other, the cyclone begins to die; for forecasting work it thus can be said, "After the occlusion the cyclone soon begins to fill up."

Structure and energy could be related to the cyclone's propagation. A young cyclone, whose center is defined as the crest of the warm tongue, propagates in the same direction as the wind in this warm sector. As the wind velocity increases within this sector, so too does the forward propagation of the entire system. Forecasters could assume that during a cyclone's early life, the system will move with accelerating velocity until the storm reaches maximum intensity, that is, until the cyclone reaches the occlusion phase. Once occluded, the cyclone moves more slowly and even becomes stationary. Next, the group sought criteria for determining the stability of a cyclonic wave: under what conditions would such a "wave" on a polar front (Fig. 23b) remain stable or, alternatively, grow in amplitude and eventually—in occlusion—become a vortex of cold air?

Calwagen's rigorously analyzed maps confirmed a suggestion by Solberg that the original single polar front actually consists of four main polar fronts circling the Northern Hemisphere. Although for theoretical purposes Bjerknes preferred to consider a single polar front to make the problem simpler, the forecasting work increasingly reinforced a more nuanced picture that gradually became Bergen school doctrine. Having installed wireless telegraphic apparatus at the weather bureau, they could readily integrate in their analyses the

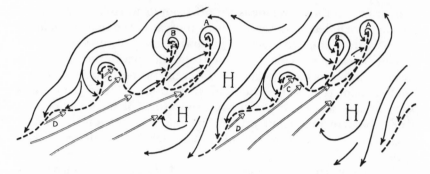

Fig. 24. Cyclone families. From J. Bjerknes and Solberg, "Life Cycle of Cyclones," p. 12.

continuing extraordinary expansion in international weather data, including data from an ever-increasing number of ships in the Atlantic. For improving the forecasting work, the Bergen meteorologists identified individual polar fronts, tagged the cyclone families found on each one, and assigned numbers for tracking movement and development of each family member. This routine enabled them to introduce a greater stability to the map analyses, which in turn allowed the entire group to agree on the nonunitary nature of the polar front around the hemisphere. The resulting schema became the basis for forecasting and for theoretical inquiry.

Each polar front section extends from southwest to northeast and moves in a general motion from west to east (Fig. 24). Cyclones originate on the southwest end of a polar front and propagate as waves along the surface of discontinuity. During the propagation toward the northeast, and wavelike cyclones gradually become, through the occlusion process, stationary vortices. The propagation velocity of an individual cyclone generally will be greater than that of the polar front, on which the "cyclone family" develops. On each polar front several cyclones at different stages of development can usually be found. The fact that each cyclone in such a "cyclone family" seems to move on a track somewhat to the south of its parent cyclone can help forecasters predict their motions. Each cyclone family itself is generally succeeded by a northerly air current with anticyclonic conditions (high-pressure systems). The Bergen meteorologists claimed that the time interval between the arrival of one cyclone family and the next is roughly five and a half days, which matched some of the climatological statistics on the occurrence of rain at several locations around the Northern Hemisphere.

The group confidently proposed a model for the general circulation of the Northern Hemisphere; it consisted of four currents moving south from the polar regions. Although originating at right angles to each other, the four currents of air turn to the right, as a consequence of the earth's rotation, and assume a greater northeastly-to-southwesterly direction. Between and above these polar currents, a poleward motion of tropical air flows in a southwest-to-northeast direction. As the four currents of polar air revolve around the hemisphere from west to east, so too revolve, on each left flank the discontinuities with their cyclone families, on each right flank an anticyclone. While granting that the regular motion and structure of these systems are broken by topographic features and by seasonal changes of thermal conditions, the regularity involved, according to Bjerknes and the others, still permits formulating long-range forecasts for up to five days in advance.[40]

The formulation of the Bergen school's evolutionary cyclone model and its four-polar-front model of hemispheric circulation, which became the capstones of the early Bergen school's endeavors, depended in general upon refinements in forecasting practices and specifically upon the occlusion process. Although the idea of a wave "rolling up" along a surface of discontinuity between two differing fluids (i.e., masses of air) had been suggested theoretically as early as the late nineteenth century by Helmholtz, among others, and had been considered by Bjerknes and his colleagues well before 1922, the actual depiction and comprehension of such a process had never before been realized or demonstrated. Bergen school articles and addresses from 1920 show a recognition that cyclones propagate in some wavelike manner along the polar front but no clear understanding of such a process. Once the life cycle of cyclones along a polar front had been described, there were still some uncertainty and disagreement even among the Bergen school's members over the notion of a cyclonic "wave."[41]

In retrospect this life cycle of cyclones may seem simple enough to have been directly deduced once the group began thinking of cyclones in terms of waves along a surface of discontinuity. Such was not the case. To define an evolutionary cycle, the occlusion process proved to be a necessary precondition. In spite of the fact that most cyclones reaching Europe are already occluded, not a hint of this process, or of the occluded front, can be found before Bergeron's insight in November 1919. Just two years later he had established occlusion as a fundamental process in the evolution of cyclones. Oc-

40. V. Bjerknes, "Annual Report 1922," to Carnegie Institution, 25 July 1922, copy, VBP; Bjerknes and Solberg, "Life Cycle of Cyclones."

41. Bergeron to Solberg, 9 June 1924, HSP.

clusion entails thinking in three dimensions and conceptualizing atmospheric processes and structures in three dimensions; three-dimensional thinking and conceptualizing only occurred after 1918. Bergeron's extraordinary ability to analyze the fine-scaled observations with a method of indirect aerology to construct three-dimensional models led to the establishment of the occlusion process and of the models of occluded, or upper, fronts. Without the detailed cloud and weather observations introduced after the war for aviation purposes, the observational basis for his analysis would have been lacking. Without the increase in the daily frequency of making observations and analyzing maps, also a consequence of aviation needs, the occluding process would certainly not have been clearly observable. Once or twice daily "pictures" of the atmosphere could not reveal these processes.

Finally, the need to understand the mechanics of cyclonic development to improve forecasting led both to refining the methods of analysis and to probing further the *sammenklapping* phenomenon. True, a theoretical account of cyclonic evolution would necessarily have had to include a mechanism for the death of cyclones; the discovery or derivation of such a process, however, was not at all accessible through theory. Bjerknes himself, denying that any theory or hypothesis underlaid the emergence of the occlusion process and life cycle of cyclones, asserted they stemmed instead from new forms of practice: "Its principle is: the most exact and most complete observations possible, and the most painstaking [*omhyggeligst*] analysis possible. . . . Should something be stated that distinguishes the Bergen meteorologists' working methods from others, it is the thorough work which is put into this analysis; where a multiplicity of details are taken into consideration that are overlooked as yet at the majority of other weather forecasting centers."[42] And this striving for greater precision arose from their attempts to assist Norway's economic livelihood: "For every year that has gone their [the Bergen meteorologists] work has constantly brought more responsibilities, following the increasing use that persons in the most differing forms of commerce, above all our large population of fishermen, make of the weather and storm predictions."[43] Likewise the drive for better forecasting also figured in replacing the unitary polar front with the four-polar-front model of the hemispheric circulation: "The fishermen along our 2000 miles [sic] long coast wait for the forecast if they shall go out or how far or where they shall go."[44]

42. "Professor Bjerknes' svar paa direktör Sandströms artikkel," *Bergens Tidende,* 29 December 1923.
43. Ibid.
44. V. Bjerknes to Robert A. Millikan, 18 February 1925, copy, VBP.

The development of the occlusion process and a life cycle of cyclones represents only part of the Bergen meteorologists' attempt during the early 1920s to integrate weather into economic operations. They were concurrently trying to rationalize (i.e., account for every occurrence of weather phenomena) "fair-weather" and in so doing established air-mass analysis.

[11]

Making Hay, Fair-Weather Flying, and the Origins of Air-Mass Analysis

THE Bergen school's idea of an air mass, seemingly a simple concept, resulted from a complex process spanning several years. By the mid-1920s the Bergen meteorologists recognized the existence of large bodies of air characterized by approximate horizontal physical homogeneity arising from the "life history" of the mass. These air masses form whenever an extensive portion of the atmosphere remains at rest for sufficiently long periods over a broad region possessing uniform surface properties. The body of air thereby attains an equilibrium with the surface, which results in a common vertical distribution of temperature and moisture over a broad horizontal area. Such regions as the snow-covered areas of Siberia in winter or the uniformly warm waters of the Gulf of Mexico can serve as "source regions" where air masses acquire distinct physical characteristics. When an air mass finally moves away from its source region and passes over surfaces with different properties, the equilibrium previously established is disrupted; the air mass's structural characteristics begin to change.[1]

Especially in its upper levels, each air mass possesses particular conservative properties that make it possible to classify its origin and trace its daily movements and physical changes. The value for weather prediction arises from the knowledge that during a given season, air masses from particular source regions possess specific characteristics that, depending on where they go after leaving the source

---

1. For example, when moving over an area warmer than its source, an air mass will be heated from below, which can make it unstable, generally producing showers.

region, will undergo predictable modifications. Predicting the characteristics of an air mass by determining its origin and trajectory would permit forecasters to predict in turn the clouds and weather that would occur in the air mass.

Although meteorologists—and the weatherwise—had of course long known that air possesses different qualities (such as warm, cold, moist, and dry) and exists in relatively homogeneous bodies or masses, these insights played little if any role in practical forecasting work. The Bergen school tried to establish an air-mass concept and a method of air-mass analysis in connection with their attempt to rationalize so-called fair-weather meteorology. They recognized the need to study weather phenomena not associated with fronts or cyclones; for weather to be a more useful resource in commercial activities, previously ignored weather conditions needed to be included in their meteorological system.

The Problem of Forecasting Local Showers

One of the difficulties encountered during the initial summer forecasting on behalf of Norwegian farmers was in predicting local afternoon showers. These appeared to spring up in a virtually random manner and to move without apparent system. The showers did not occur in conjunction with fronts or cyclones; they often formed on what had begun as clear, sunny days. Why did showers form on some fair days and not on others? Why in some locations and not in others? Might it be possible at least to forecast when, in spite of a brilliant sunny morning, showers might occur and when they were not likely to? As Bjerknes noted,

> Everybody is acquainted with the summer days which begin with a fine morning, but as the sun rises the sky becomes cloudy, and in the afternoon suddenly a shower comes, often with thunder. In the evening it clears again, and a bright morning follows. But as the next day advances the events repeat themselves; and in this way it may continue with showers in the afternoon day after day, to the despair of the peasant during the hay-harvest. These showers seem more mysterious the more closely we try to examine the conditions connected with their appearance.[2]

Meteorologists had traditionally ignored the problem of forecasting local showers, which were considered to be among the most random of

2. V. Bjerknes, "The Structure of the Atmosphere When Rain Is Falling," *Q JRMS* 46 (1920), 131.

weather phenomena.[3] Neither the economic and social consequences
of local showers nor the scientific and technological resources available
to forecasters encouraged meteorologists to study or predict this phenomenon in detail. Two needs, however, made this problem acute for
the Bergen meteorologists. In common with meteorologists in other
nations who were concerned with weather forecasting for aviation,
showers and local thundershowers required attention because of the
real danger they posed for aviators. As Bjerknes noted from the start,
"Warnings of such showers along aerial routes will therefore become a
need that cannot be ignored. A forthcoming task will be to organize a
weather service that can take care of this [problem]."[4] The other need
was much more immediately pressing and much more a specific Norwegian problem at that time: to make hay.

Normally haymaking was not a major concern of forecasters, but
during the first years of the Bergen school, several factors kept it in
focus. Government intervention to increase the production of grains
caused hay to become scarcer and costlier.[5] Moreover, after the war
the net income for farms began to drop substantially; labor costs,
which continued to increase until 1921, remained higher than the
falling gross income for farms.[6] Bjerknes understood the significance
for the farmers of accurate predictions of local fair-weather showers:

> The weather's caprices result in carrying out an enormous amount of
> unnecessary work during the sowing and harvesting seasons: the mow-
> ing of hay or the cutting of grains on an ill-timed day; the hay is tossed
> out only immediately afterwards to have to be raked together again, etc.
> A great part of this work should be able to be saved by reliable forecast-
> ing. . . . The greatest number of erroneous forecasts are caused by the
> capricious local showers which during the summer spring up . . . and
> which produce so many disturbances during haying season.[7]

3. William Napier Shaw, *Forecasting Weather* (London, 1911), pp. 295–96, discusses
the frustrations of trying to predict such showers. He resorts to the expression that in
such periods the weather has "moods" or "fits." "Rain sometimes falls without any
recognisable meteorological reason, even, as we have seen, in the central region of an
anticyclone."

4. V. Bjerknes, "Om forutsigelse av regn," *Naturen* 43 (1919), 345–46.

5. Wilhelm Keilhau, *Norge og verdenskrigen*, Carnegie Endowment for International
Peace: Economic and Social History of the World War, Scandinavian Series (Oslo,
1927), p. 270, claims that increased production of grain was at the cost of reduction in
hay. He provides the following annual production figures for grain and hay respectively (in tons): *1916*, 351,598 / 3,075,007; *1918*, 440,694 / 1,685,701; *1921*, 352,472 /
1,834,928. Hay prices received on the Christiania market are found in *NOS XII 245*,
Table 272. In Crowns per 100 kg., timothy hay increased from 13.58 in 1916 to 29.09
in 1920; meadow hay increased correspondingly from 11.00 to 25.76.

6. *NOS XII 245*, Table 100, "Working Results of Farms" and Table 292, "Wages in
Agriculture and Forestry."

7. V. Bjerknes, "Om vær- og stormvarslinger og veien til at forbedre dem," *Teknisk
Ukeblad* 38 (1920), 300, 302.

In haying season, Bjerknes noted, using the correct days properly could be decisive; otherwise, "the hay can rot in the haystack and even dried hay can easily lose half its nutritional value during a rainy period."[8] Not only could accurate forecasting of showers (and of more general rain) for specific regions save the important hay harvest; it could help the farmers plan and organize their work efficiently. An assurance that the weather would remain dry for several days could permit labor-saving methods of drying hay by tossing it on a hillside (*bakketørking*) rather than drying it on a rack (*hesjetørking*).[9] Repeatedly, Bjerknes emphasized the potential savings from organizing farm work with the aid of reliable forecasts. At a time of increasing government subsidies for agriculture and of declining profitability, reliable weather forecasts could help rationalize production.[10]

From 1918 to 1920, farmers were the most important social group whose support Bjerknes needed for maintaining the expanded forecasting services; their belief in the benefits of forecasting to them was essential. The Bergen meteorologists thus felt compelled to investigate the nature of local showers and to find criteria for predicting them. Whereas Jacob, for example, originally opposed devoting time to what he called, "fair-weather meteorology," believing that fronts were all that needed to be studied, he eventually came to realize the political consequences of losing the support of farmers.[11]

An early indication of the importance of local showers came in a letter from one of the forecasting service's observers. After reading in the newspapers that the statistics for the summer 1918 experiment were rather favorable, Sigurd Enebo, an astronomer and weather observer for the forecasting service, wrote to the Meteorological Institute highlighting some discrepancies between the published conclusions and his own official weather statistics for Dombås, at the northern end of the fertile Gudbrandsdal.[12] He pointed out that the majority of the erroneous forecasts entailed rain on days that dry weather was predicted. This type of error causes more problems for

8. V. Bjerknes, "Værvarsling," in series, *Universitetets radioforedrag* (Oslo, 1933), p. 63.

9. V. Bjerknes, "Om vær- og stormvarslinger," p. 300; "Værvarsling," p. 63.

10. Toward this end the newly founded Central Committee for Scientific Cooperation for the Advancement of Industry established a subcommittee for studying the means by which meteorology and climatology could be used to aid agriculture. Bjerknes worked with this committee in 1920 and continued to follow the effectiveness of the forecasting services for the farmers and for the various efforts toward improving agricultural production.

11. Related by Bergeron, untitled MS on events from 1919–20, 1929, TBP; see also Theodore Hesselberg, "Beretning om den utvidede veirtjeneste for jordbruket paa østlandet," 10 December 1918, KUD/D, folder: "Veirvarslinga på Vestlandet," file: "Budsjetter 1918/1919–1923/1924"; NMIÅ *1919–1920*, pp. 19–20, 27–28.

12. Sigurd Enebo to Hesselberg, passed on to H. Solberg, 2 November 1918, HSP.

farmers than dry weather when rain is predicted; to make matters worse, most of these errors occurred in the most hectic time of the summer, August. Trying to take the sting out of this bad news, he volunteered, "In the warm period it is almost impossible to avoid the small (or larger) local showers." After describing the typical cycle of cumulus and cumulonimbus clouds forming on days that begin sunny, turn to rain showering down in various amounts by afternoon, and end by clearing, he conceded that in such periods, "the weather is most difficult to predict here at the same time that to know it is pressing [*magtpaaliggende*] . . . of course I am not here saying that the summer weather forecasts didn't have their value. And I hope and believe that in the future they will have great significance for the farmers."[13]

The problem received some attention during the 1918–19 winter.[14] By studying the detailed maps from the summer forecasting, Solberg and the others found specific regions where showers tended to form. Emanating from these so-called centers of action, the showers then drifted away from their points of origin. These charts, because they revealed where summer showers were most likely to form and how they tended to migrate, could be useful in planning future flying routes.[15] The investigation did not, however, resolve the puzzle of why the showers form on one particular fair day and not on another, although it did produce some clues.

Showers seemed to occur when conditions existed that allowed moisture-laden sea breezes to flow far inland. Apparently topographic features of the landscape forced the moisture-laden breezes to converge toward a point, which in turn caused the air to ascend. When rising air cools below its saturation point, clouds and showers form. At this time the Bergen group still used the kinematics of wind flow as the basic referent for developing explanatory models; convergence points in the wind flow, where the moist air was forced to rise, were associated with the "centers of action" where showers tended to form. The group concluded that daily weather maps showing the distribution of wind and of the air's water content (absolute humidity) would

13. Ibid.
14. V. Bjerknes, "Forslag til forbedring av veirvarslingen for landbruket sommeren 1919"; J. Bjerknes and H. Solberg, "Nogen resultater av veirvarslingen for landmænd sommeren 1918"; T. Hesselberg, "Beretning om den utvidede veirtjeneste for jordbruket paa østlandet sommeren 1918." All sent to KUD, 10 December 1918, KUD/D, "VpV"/"Budsjetter 1918/19–1923/24;" V. Bjerknes, "Om forutsingelse av regn," 336–45.
15. Flyveskolen to A/S Nordisk Luft Kraft, 3 September 1919; Nordisk Luft Kraft to NMI, 4 September 1919; both in NMI; V. Bjerknes, "Om forutsigelse av regn," pp. 345–46.

be of greatest value for predicting when showers would most likely occur in the vicinity of these centers of action. The arrival inland of moisture-laden air by the action of diurnal sea breezes would be the primary consideration for predicting the probability of local summer showers on an otherwise sunny day. The next summer forecasting period would provide opportunity to apply this insight and learn more.

Having returned to Christiania for the summer 1919 forecasting, Solberg analyzed the shower problem carefully during the course of his forecasting work. While preparing forecasts for East Norway's agricultural regions and for various air bases, he recognized that the preliminary explanation for the formation of these local summer showers was inadequate. In a letter to Jacob Bjerknes, he wrote, "In the course of the forecasting I have gotten the impression that we have not found the entire truth during the winter. In order that local showers occur [moist] sea air is, to be sure, a necessary condition, but unfortunately not a sufficient one, speaking mathematically. But now I have good hope indeed to find also this sufficient condition."[16] In Bergen meanwhile, Bergeron began to interest himself in this problem as well. He informed a Swedish acquaintance that the Bergen group had come across the secrets of local showers; rather than being random phenomena, they actually follow "the laws of nature."[17] But a solution to the shower problem was not immediately found.

The Bergen meteorologists recognized that the occurrence or non-occurrence of local showers did not depend on the pressure configuration over the region.[18] Analyses of the situations in which showery periods developed or ceased revealed that the pressure distribution was not predictive of these events. The traditional identification of fair weather with areas of "high pressure" (anticyclones) could not at all be relied on: "Such striking contrasts of weather at nearly the same distribution of pressure are frequently experienced, and bring much trouble for forecasters."[19] Development of showers on what otherwise would be fair-weather days seemed to depend both on the moisture content of the air, as they had recognized during the winter of 1918–19, and on the relative vertical stability of the air.

In simplest terms, vertical stability is the air's ability to produce and

16. Solberg to J. Bjerknes, 22 August 1919, JBP.

17. Tor Bergeron to Gunnar Rising, 20 July 1919, copy, TBP; Bergeron to Alex Stafwerfeldt, 15 August 1919, copy, TBP.

18. Determining who actually carried out the investigation is difficult; in all likelihood Bergeron was involved as well.

19. J. Bjerknes and H. Solberg, "Meteorological Conditions for Rain," *GP* 2, no. 3 (1921), 51, 36–60.

to sustain vertical currents. When the sun-heated ground warms the air closest to it, parcels or bubbles of heated air rise and, in so doing, begin to cool. In a stable situation the rising air parcels soon find themselves cooler and denser than the surrounding air and therefore tend to sink down. In unstable air, the rising heated air will be warmer and less dense than the air surrounding it at each level of its ascent; hence buoyancy forces support continued vertical displacement upward. If the rising air contains sufficient moisture and if it is cooled to the point that condensation occurs, cumulus-family clouds and ultimately showers may form. If the mass of air is very unstable, allowing strong vertical currents to develop, thunderstorms may result.

During the summer, for example, warm air arriving in Norway from the south or southwest may be cooled in its lowest layers while traveling over colder sea surfaces, but higher up the original warm characteristics remain. This stable configuration with dense, cold air at the bottom inhibits convection; therefore, even when carrying considerable moisture such stable air will not produce showers. Cold polar air arriving over Norway in the summer will be heated by the warm ground below. In spite of a situation that favors vertical motions, such air generally contains little moisture so showers do not develop. When unstable polar air of this sort *does* contain sufficient moisture, as when local sea breezes bring moisture far enough inland, then shower formation becomes likely. This explanation accounted for situations in which, under similar pressure distribution over several days, afternoon local showers did not at first occur but after one or two days became increasingly more prevalent. Unstable, dry polar air over Norway, being heated from below during sunny days, only gradually received enough water vapor from diurnal sea breezes to provide the rising air currents with sufficient moisture to condense and form showers.[20]

Jacob Bjerknes and Solberg thus concluded that "the properties of the air in respect to stability and content of moisture are more important factors for the occurrence or nonoccurrence of local showers than the general distribution of pressure."[21] If stability and moisture content were to serve as criteria for predicting local summer showers, however, methods to use them in practice would be necessary.

The Rationalization of Fair-Weather and Air-Mass Analysis

Concurrent investigations into means of reproducing fronts in forecasting practice had revealed that a body of air's past history

20. J. Bjerknes and Solberg, "Meteorological Conditions for Rain," pp. 36–37ff.
21. Ibid., p. 59.

largely determined its physical properties. This new insight helped the Bergen school formulate its air-mass concept. While attempting to find methods of locating and identifying cyclonic fronts and the polar front, Bergen meteorologists had noted that the air's past movements seemed to determine many of its physical properties. In polar air, temperatures near the surface could be altered when the air moves for a period of time southward over warm water, thereby masking near the surface any thermal discontinuity between air of polar origin and physically different air of tropical origin. Such difficulties in reproducing fronts in practice forced the Bergen meteorologists to examine more carefully their means of distinguishing between the air on either side of a discontinuity.

Consequently, they found "the thermal boundary line is no longer recognized only by the jumps in temperature—this can easily be more or less blurred as a result of local warming or cooling—but on a number of other signs that indicate air of different origin."[22] At first they used discontinuities in pressure, wind, and humidity as well as cloud types and the limits of rain areas as indicators of the existence of a front. During the fall of 1919, Bergeron's visibility studies revealed that the air's transparency is a sufficiently conservative property that it could be used to distinguish air of different origin, not unlike using staining liquids to trace movements.[23] In working on methods for locating fronts they increasingly used an implicit and eventually explicit understanding that the characteristics of bodies of air depended on the past movements of the air.

This notion was then assimilated into the efforts to resolve the shower problem. Recognizing that the pressure distribution provided little assistance in forecasting showers, the Bergen group turned increasingly to the life history of air masses as a means of accounting for vertical stability and perhaps predicting the occurrence of local showers. Being unable to rely on such direct measurements, they proposed using indirect signs based on clouds and weather phenomena to infer the stability and origins of air masses.

These investigations were at the same time aimed at finding a means of identifying the boundaries of air masses in situations in which locating the polar front was otherwise impossible. For example, polar air masses having moved southward can be warmed from be-

22. V. Bjerknes, "Om vær- og stormvarslinger," p. 304.

23. Bergeron discovered that a "remarkable opacity appears suddenly whenever the air of the 'warm sector' of an atlantic cyclone invades the space of observation. A most remarkable coincidence between the disappearance of the opacity and the squall line [cold front] also occurred several times"; Bergeron to William Napier Shaw, 18 December 1919, copy, VpVA; see also Bergeron to Arvid Carlstedt, 15 October 1919, copy, TBP.

low. Convective currents produced in this manner produce cumulus clouds and even cumulonimbus clouds accompanied by local showers. Although temperatures near the surface may be as warm as those in a neighboring air mass of nonpolar origin, the occurrence of cumulus and cumulonimbus clouds and local showers solely within the polar air mass can reveal the boundary between the physically different air masses. Similarly, the appearance of stratus-family clouds in an air mass can reveal air of tropical origins that has been cooled in its lowest layers while moving northward in the temperate zone. The low-level stratified clouds signify the stability of a warm air mass that has been cooled from below in its lowest layers; the horizontal extent of such clouds could reveal the boundaries of this air mass.[24]

Thus by 1922 the Bergen group clearly recognized and could distinguish large, distinct air masses, defined physically by their life histories. The concept at first had only limited use as a prognostic technique. The all-important air-mass characteristic, vertical stability, is essentially aerologically defined. Indirect aerology could provide only general qualitative indications of the stability of an air mass; quantitative measurements of the vertical temperature distribution would be necessary to forecast detailed meteorological conditions within an air mass, in other words, to define criteria for classifying and identifying air masses.

When Calwagen began working in Bergen he too turned to the problem of local showers and air masses; when he became head of the Bergen Observatory in 1922, he initiated a program of obtaining direct aerological measurements in an effort to predict showers. His interest entailed weather phenomena not associated with cyclonic and frontal activity. This work had its origins partly in the original problem of local showers and partly in attempts to improve aviation forecasts. Experience had shown that fair-weather flying had its share of meteorological problems.

Both military and commercial flights in Norway had given meteorologists in Bergen and Christiania opportunities to experiment with forecasting for aviation. From mid-August to mid-October 1920, Det Norske Luftfartrederi had operated an experimental Bergen–Haugesund–Stavanger air-mail and passenger service along the southwest coast.[25] The meteorologists in Bergen had worked closely with the aerial service, providing forecasts by telephone for several

24. J. Bjerknes and Solberg, "Life Cycle of Cyclones and Polar Front Theory of Atmospheric Circulation," *GP* 3, no. 1 (1922), 3–8.

25. See Nordic Pool for Aviation Insurance, *Luftfartsbok omhandlende Danmark, Finland, Norge og Sverige* (n.p., 1922), pp. 154–66, for technical, economic, and operational details.

hours in advance and for the next twenty-four hour period.[26] Additional experience had come from close contact with the army and navy's flying operations in both inland and coastal regions.[27] Although Norway's commercial aviation corporations were liquidated when the depression began, interest and activities in aeronautics had not ceased.[28] Even if Norwegian civil aviation had to wait for better economic times, military aviation continued to require weather services. The head of the army's air force claimed that "a modern air force cannot manage without a well-arranged weather service."[29]

In 1922 the Norwegian weather services reported that they would "in the coming year, so far as possible, continue work to get a solid foundation for an aviation weather service."[30] Even from the few Norwegian experiences, no less from the voluminous reports in the international journals, additional unforseen problems had arisen. Visibility, turbulence, strong gusty winds, and showers not related to fronts and cyclonic activity had proved troublesome during the Norwegian flights in 1920.[31] The need to issue forecasts along a flight route that could provide information on problematic fair-weather phenomena informed Calwagen's research program—a research program that "has first and foremost taken aim at the possibility of improving and developing the forecasting work."[32]

26. Forsvarsdepartementets luftfartsråd to Forsvarsdepartementet, 13 June 1922, on weather services for aviation in Norway, copy, KUD/D, "VpV"/"Budsjetter 1918/19–1923/24."

27. See Nordic Pool, *Luftfartsbok*, pp. 167–72, passim, for details on Norwegian military flying operations; Marinens flyvevesen to Forsvarsdepartementets luftfartsråd, 8 May 1922, copy; Våbeninspektøren for hærens flyvesen to Forsvarsdept.'s luftfartsråd, 11 May 1922, copy, on the meteorological needs of the navy and army air forces and the assistance received from the weather services, KUD/D, "VpV"/"Budsjetter 1918/19–1923/24." VpV in Bergen was called upon to assist with the forecasts for specific operations, even when these were in districts normally falling in NMI Christiania's jurisdiction.

28. Nordic Pool, *Luftfartsbok*, p. 166; plans existed for an airship route, London–Christiania–Stockholm–Copenhagen–London, H. Riiser-Larsen, *Femti år for Kongen* (Oslo, 1957), pp. 84–90.

29. Våbeninspektøren to Forsvarsdept.'s luftfartsråd, 11 May 1922; Hærens flyvevesen and Marinens flyvevesen to Forsvarsdept.'s luftfartsråd, 11 May 1922, 8 May 1922.

30. "Veirtjeneste for lufttrafikken," in Nordic Pool, *Luftfartsbok*, pp. 178–79.

31. Ibid., p. 162, passim.

32. Ernst G. Calwagen, "Plan for de aerologiske arbeider ved Meteorologisk Observatorium Bergen," esp. section "Utdrag av betænkning vedrørende Observatoriets aerologiske arbeider" (n.d. [but easily dated 1922 from internal evidence]), KUD/D, "VpV"/"Budsjettsaker 1916–46." When Calwagen became director of the observatory in 1922 after B. J. Birkeland, he embarked on a plan to expand and strengthen the aerological research that had previously been pursued there. In this manner Calwagen, who had been affiliated with VpV, helped integrate the observatory into the Bergen school's domain.

"To gain experience in forecast work for aviation," Calwagen ventured to analyze and predict the properties of air masses.[33] He endeavored to provide detailed predictions for aviators of the expected
types of clouds, visibility (including fog), turbulence, and probability
of local showers or thunderstorms that might be encountered along a
flight.[34] Following up Jacob's and Solberg's conclusion that occurrence or nonoccurrence of local showers owed more to the origin and
life history of air masses than to pressure distribution, Calwagen extended the investigation to encompass also these other fair-weather
phenomena. Although acknowledging that the prediction of local
showers in the past was at best "tacitly passed over . . . as rather hopeless," Calwagen believed that with direct aerological measurement of
air masses, such predictions could be feasible. "The instability in the
vertical direction in air masses up to four or five km. in height . . . is
[the] cause of shower phenomena. Forecasting of showers and thunder[showers] has up to now always rightfully been regarded as by far
one of the most difficult problems. . . . It is in the nature of the whole
problem that the way to better predictions here comes only by purely
aerological studies."[35]

To obtain vertical distributions of several meteorological elements,
Calwagen turned to the most recent aerological instrument available,
the airplane. Toward the end of the war and during the postwar
years, meteorologists in several countries began designing instruments with which airplanes could be equipped to make aerological
soundings. Interest by the Norwegian military air forces in weather
forecasts enabled Calwagen to recruit their support for his study.[36]
He combined these aerological studies of air masses with actual forecasting work for aviation. Working primarily from the army's air
force base at Kjeller (February to March and July to August 1923;
July to August 1924) and the navy's base at Horten (September 1924),
Calwagen made regular ascents in instrument-laden airplanes. On
these ascents he recorded the vertical temperature gradient, humidity, and pressure while also observing and photographing the cloud
systems: their orientation, dimensions, motions, growth or decay, and
changes. By analyzing these observations in relation to each other and

33. NMIÅ *1923–1924*, p. 38; see also Calwagen, "Plan for de aerologiske arbeider."
34. Ernst G. Calwagen, "Zur Diagnose und Prognose lokaler Sommerschauer: Aerologische Flugzeugaufstiege in Ostnorwegen," *GP* 3, no. 10 (1926), 95–103.
35. Calwagen, "Utdrag av betænkning vedrørende," in "Plan for de aerologiske
arbeider."
36. See the various letters from the respective air forces to the Ministry of Defense's
Advisory Board for Aviation contained in Forsvarsdepartementets luftfartsråd communication to KUD, May 1922; NMIÅ *1923–1924*, pp. 38–39, and NMIÅ *1924–1925*,
pp. 69–71, on the army and navy air forces' assistance to Calwagen.

also to detailed weather maps, Calwagen issued special weather forecasts twice daily to the military air forces.[37]

In his study, Calwagen first described the meteorological conditions occurring during each airplane ascent, including the general weather patterns over northern Europe and weather, clouds, turbulence, and visibility encountered during the flight. He then diagnosed each case by air-mass analysis. He began determining quantitative criteria for identifying specific types of air masses—polar continental, polar maritime, tropical continental, and tropical maritime—and for reconstructing their trajectories from the locations of their origin.[38] Such reconstructions of the "life history" of an air mass facilitated prediction of the latter's properties upon arrival at a new location.[39] In this manner he attempted predictions not only of local showers but of other atmospheric phenomena related to the air mass's stability, such as turbulence, gustiness, and visibility.

Calwagen's work showed much promise. While making aerological soundings with an airplane on 10 August 1925, he and the pilot were killed when the plane fell apart in midair.[40] Most of his results were later incorporated into the work of his friend and colleague, Bergeron, and then overshadowed by the comprehensive endeavors in airmass analysis of the German "Bergen-ite," Gerhard Schintze, and the early American adherents of Bergen meteorology, Hurd C. Willett and Jerome Namias.[41] In the 1930s the availability of new technologies for aerological soundings led to an extensive elaboration of methods for air-mass analysis, and for that matter, for frontal

37. NMIÅ *1924–1925*, p. 68.

38. Based on Calwagen's earlier unpublished inquiry, Bergeron had adopted a classification system of four air-mass categories in his study, with Gustav Swoboda, of a stationary polar front over Europe in October 1923. In that study, they identified different air masses and the stability criteria for waves on the polar front. T. Bergeron and G. Swoboda, "Wellen und Wirbel an einer quasistationaren Grenzfläche über Europa," *VGL* 3, no. 2 (1924), 63–172. Bergeron repeatedly advised me not to forget Calwagen's work: "A history of air-mass analysis must not omit Calwagen's pioneering study"; interviews with Tor Bergeron, Uppsala, February and June 1976.

39. Calwagen, "Diagnose und Prognose lokaler Sommerschauer," pp. 33–40, 51–63, 91–96, and esp. the detailed analyses, "Diskussion der Einzelfälle," 64–90.

40. NMIÅ *1924–1925*, p. 68.

41. Hurd C. Willett, "North American Air Mass Properties," *Journal of Aeronautical Science* 1, no. 2 (1934), 78; and reprinted in the many editions of Jerome Namias's *An Introduction to the Study of Air Mass Analysis* appearing in the 1930s. The new American meteorological school centered around C.-G. Rossby at M.I.T. picked up and elaborated air-mass analysis in the late 1920s and 1930s. Because of its significance for aviation weather forecasting, air-mass analysis naturally became a subject for the newly founded (1928) M.I.T. meteorological section (later department), which was established with aid from the Daniel Guggenheim Fund for the Promotion of Aeronautics. See Richard P. Hallion, *Legacy of Flight: The Guggenheim Contribution to American Aviation* (Seattle and London, 1977), pp. 220–21.

analysis and cyclone theory. Most notable was the radiosonde; sent aloft by small balloons, it radioed back to earth measurements of pressure, temperature, and humidity as the balloon and instruments ascended. Although more powerful and sophisticated than the early 1920s air-mass analysis, the mature methods rested on the earlier Bergen studies.

Air-mass analysis expanded the utility of weather as a resource in commercial ventures. Although the particular problems of Norwegian agriculture in the early 1920s made local showers an especially pressing issue for Norwegian meteorologists, aviation's needs reinforced their drive to understand and predict such showers, and other previously ignored non–front-related fair-weather phenomena as well.

Conclusion

VILHELM BJERKNES made history, but not the history of his choosing. His career developed in a manner he had never envisioned; so did the science of the atmosphere he endeavored to shape. Both his professional evolution and the science he established were shaped by unexpected exigencies. He learned early that curiosity, vision, and innovative work were not sufficient to secure success in professional science. Success would depend as well on convincing other scientists to adopt his research problems and methods and on placing his disciples in authoritative situations where their reputations could contribute to both his prestige and his program.

Bjerknes's career reveals how necessity led an astute scientist to define and transform research programs, disciplinary agendas, and criteria for constituting knowledge. Bjerknes the strategist was guided by his understanding of actual and potential audiences and resources. That is, he grasped the outlines of a political economy of institutionalized science and adapted his strategies to the ecological relations within and among disciplines.[1] He assessed what institutions and research programs might draw support; what scientific and societal constituencies might make use of his work; what employment pos-

1. Charles E. Rosenberg, "Toward an Ecology of Knowledge: On Discipline, Context, and History," in *The Organization of Knowledge in Modern America, 1860–1920,* ed. Alexandra Oleson and John Voss (Baltimore and London, 1979), 440–51; Robert E. Kohler, *From Medical Chemistry to Biochemistry: The Making of a Biomedical Discipline* (Cambridge and New York, 1982); Robert H. Kargon, *Science in Victorian Manchester: Enterprise and Expertise* (Baltimore and London, 1977); Pierre Bourdieu, "The Specificity of the Scientific Field and the Social Conditions of the Progress of Reason," *Social Science Information* 14 (1975), 19–47.

sibilities existed for those he might train; how he might compete for resources and authority with which to shape, produce, and circulate knowledge; and whether his endeavors would bring him prestige. He acted at times with clear, well-thought-out strategies for exploiting existing or anticipated conditions, at other times, seemingly instinctively, responding to immediate impressions without regard to long-term consequences.

Bjerknes's professional options in physics were limited by both his disposition and his circumstances. Stockholm was a poor location for participating directly in the ferment of turn-of-the-century physics. At first he thought his research program was one that would remain central to European physics and could be his vehicle to prestige and authority, but he was mistaken. In Sweden there was little sympathy for theoretical physics. Moreover, the impoverished Stockholm Högskola lacked facilities for experimental research, and the two older universities controlled most of the scarce funds, students, and employment situations.

Bjerknes took a calculated risk and transferred his mechanical physics to the atmosphere. Although nobody in the early 1900s could foresee how physics, meteorology, and organized flight would develop, Bjerknes assumed that the aeronautical conquest of the air would both depend on and contribute to the scientific conquest of the weather. This idea of a physics of the atmosphere was informed not only by his desire to maintain a mechanical world view and by his understanding of the developing technological foundation for such a science, but also by an adaptive professional strategy. He recognized the obstacles to gaining a foothold in meteorology proper, which was already institutionalized in Norway and Sweden, so he chose a position that did not immediately threaten established interests, and attempted to create a niche that might at first coexist with the established professional structures and draw upon alternative resources and audiences. To the extent that Bjerknes could have a clear strategy during this period of great change in physics and atmospheric science, he seems to have conceived of transforming aerology into a mechanical physics of atmospheric changes; for within atmospheric science, aerology, that "menagerie . . . of dilettantes,"[2] seemed to him to have the greatest potential for, and fewest institutional obstacles to, growth as a discipline under his influence. Bjerknes believed his physics of the atmosphere could eventually become the scientific basis for a rigorous, professional aerology, but first the science had to be established.

2. Bjerknes to Svante Arrhenius, 4 March 1909, SAP.

Bjerknes soon realized he needed data for his project which only the international community of aerologists, with its budding aeronautical meteorological concerns, could provide. His failure to persuade this community to adopt his system of aerological measurements and rational methods led him in turn to realize the necessity of producing a new generation of aerologists with which to establish his new discipline; but this he could not do in Christiania. The political economy of science in Wilhelmian Germany provided the opportunity through its need for scientifically trained aerologists for military and commercial flight. Amply supported in Leipzig, Bjerknes began to claim authority for himself, his project, and his institute. Without changing his long-term goals or underlying assumptions, he tried to show that issues arising from flight could best be treated within the framework of his project. Then came the war.

To continue his project in Bergen, Bjerknes had to adapt once more to local conditions. He tried to capitalize on wartime exigencies by proposing first a field weather service for the military and then an emergency weather service to assist agriculture, in both cases with an eye toward the opportunities commercial aviation would create for advancing his discipline at war's end. His goal was to assure his influence on international meteorology by developing methods in Norway that could meet postwar challenges and at the same time gain acceptance as more scientifically legitimate than existing methods.

Local conditions again forced a strategy change after the war. The new university in Bergen was not to be built; the summer 1918 forecasting experiment had showed that neither the methods derived from his physics of the atmosphere nor the state of aerological technology could satisfy local and international needs. Bjerknes saw that his best chance now lay in establishing a permanent practical forecasting service and defining new predictive methods. He set out to integrate an academic research program with a government service-oriented institution. Toward this end he once more adapted his project for creating an atmospheric physics: indirect methods of applying physics in practical forecasting would provide new insights while addressing political-economic concerns, especially the needs of aviation.

At the same time, Bjerknes sought an institutional springboard to national and international influence. He saw that weather forecasting must become a sophisticated scientific profession that included research as an integral part of the job. To the government he recommended a fusion of theory and practice in a weather service in which forecasters would pursue research on problems arising during operational work and increased salaries would attract university science graduates. This merger of theory and practice to the benefit of both seemed the only way to create a school that could realize his vision in post–World War I Norway.[3]

During the early years Bjerknes and his school ably managed to combine the search to know with the imperative to serve public interests. Conceptual change and new insight into atmospheric processes resulted from demands on, and transformation in, practice. The Bergen group hoped the weather service could continue to serve as a hothouse for deeper inquiry into and understanding of atmospheric processes, but although the ideal of forecasting as experiment and inquiry united with service to commerce and the public was attainable, maintaining a balance over time proved difficult. Bergeron warned his colleagues against "creeping empiricism": even their own methods could degenerate into mechanically performed routine, which would produce no insights and make forecasting less attractive to talented scientists.[4]

By the mid-1920s Bjerknes and his Bergen school were well on the way to appropriating the weather as professional property. At home they managed to secure their institution's future. To reduce government expenditure the Ministry of Church and Education in 1924 recommended the consolidation of the two weather centers for southern Norway. Meteorologists could choose which to close, and the other would become the new Norwegian Central Meteorological Institute.[5] Meteorologists in the capital claimed their center should be retained because the nation's capital should have the weather service for aviation. In Bergen and along Norway's west coast, shipping organizations, harbor authorities, local chambers of commerce, and most of all, fishermen argued that their weather center must remain open. Fishermen asserted that the introduction of weather forecasting was the best thing the state had done for them. Local newspapers and Storting members joined the campaign. In the end, both bureaus remained open.

Bjerknes was delighted. Fishermen and others had acknowledged the important contribution of science to their lives.[6] Although far from infallible, the new forecasting methods had proved so satisfacto-

---

3. Another Norwegian research program that attained international significance at this time apparently was established and sustained by a similar integration of a practical service role and academic scientific interests: V. M. Goldschmidt's geochemical-mineralogical studies benefited from the State Raw Materials Laboratory that was connected with the otherwise impoverished university mineralogical institute in Christiania.

4. Tor Bergeron to Halvor Solberg, 7 June 1924, HSP.

5. NMIÅ *1923–1924*, pp. 3–6; NMIÅ *1924–1925*, pp. 3–7.

6. Bjerknes to Robert S. Woodward, 7 July 1924; Bjerknes to Robert A. Millikan, 5 January 1925; copies of both in VBP.

ry that the initially skeptical fishermen increasingly abandoned their folklore and turned to the Bergen forecasts for their weather information. Fishermen would no longer comprehend the weather solely through their own direct experience, language, and prognostic signs. The weather would primarily be mediated for them through the scientists sitting in Bergen. Other sectors of society also began to understand that weather forecasts could contribute to efficient or safe operations. Weather was no longer a capricious phenomenon; information on it was becoming a resource for rationalizing commercial activities. For the meteorologists this resource was a commodity that could be exchanged for institutional support.

Internationally, Bjerknes and his school began to assume greater authority. More and more meteorologists and institutions, recognizing a need to reform their forecasting methods and research orientation, turned to Bergen for assistance. Although widespread acceptance of Bergen meteorology was yet to come, Bjerknes detected a mass psychology operating in the process of acceptance of the polar front meteorology: as soon as several meteorologists had begun claiming that they too had "seen" polar fronts, virtually everyone began to see them. The Bjerknes school's results were no longer being dismissed as "Bergen haughtiness [*Hochmut*]."[7] Although much remained to be done, Bjerknes understood at this point that he had already made his mark in science.

By the 1930s short-sighted politicians and brutal economic realities quashed the opportunity for combining practical service roles with academically satisfying work to the benefit of culture, state, and society. Budget restrictions limited the number of meteorologists hired; constant stress and routine work inhibited inquiry. Fewer positions meant all available time had to be devoted to forecasting. Both abroad and eventually in Norway, the needs of aviation began to hinder the creative union of research and prediction. Weather services for aviation no longer had time for theoretical problems arising from operational work.[8] No doubt, too, the special institutional circumstances and esprit de corps in Bergen at the start, including Bjerknes's charismatic informal leadership, could neither be maintained indefinitely nor easily transplanted.

Although Bjerknes's attempt to bring the seminar and the research laboratory into government weather forecasting bureaus enjoyed only short-term success as an institutional strategy, his claim that the Bergen school's methods were scientific, based on physical principles,

7. Bjerknes to C. W. Oseen, 26 October 1926, CWO.
8. Bjerknes and Hugo Hergesell had already discussed aviation's negative effects on meteorology and their disappointment in their correspondence during the mid-1920s.

and in direct contact with research and theoretical inquiry enabled meteorology to claim greater legitimacy as an academic science. Although academic positions created for training forecasters also made possible the institutionalization of theoretical research, a new division between theory and practice grew as university degree programs and departments of meteorology began to flourish in Europe, Japan, and North America. Bjerknes eventually abandoned the Bergen ideal and sought greater government support for "pure" research at the university in Oslo (where he returned in 1926) by stressing the eventual social benefit.[9] But the political economy of Norwegian science between the wars could support only limited academic research and only limited employment opportunities for meteorologists. Perhaps in response to the frustrations of economically strangled institutions, Bjerknes exclaimed in a moment of despair that whereas Norway might have developed "a meteorological élite," it had instead developed, "the word is ugly, but I have to use it, a meteorological proletariat."[10]

The history of science has only begun to explore the evolution of organizational forms for research and of the rhetoric by which scientists justify such structures.[11] A sharp division of scientific labor into "pure" and "applied" is of course just one choice of how to organize research, and not the only one that produces first-rate science, as the early years of the Bergen school demonstrate. Bjerknes might well have preferred an academic environment that left him free of the need to justify his endeavors to politicians and interest groups, but the combination of his professional aspirations and the existing resources made such a choice impossible. Nor is it likely science or society would have been so well served had circumstances permitted him to establish a physics of the atmosphere as a problem divorced from thoughts

9. Bjerknes wrote many historical and science policy articles during the 1930s that call for increased government support of university research as a means to combat unemployment, economic depression, and cultural backwardness.

10. Bjerknes to Bjørn Helland-Hansen, 18 July 1937, copy, VBP.

11. On how researchers in corporate laboratories requiring short-term payoffs or in bureaucratized government service institutions confront the problems of combining a variety of goals and interests in a research program, see, among many publications, George Wise, "Ionists in Industry: Physical Chemistry at General Electric," *Isis* 74 (1983), 7–21; Lenard Reich, "Irving Langmuir and the Pursuit of Science and Technology in the Corporate Environment," *Technology and Culture* 24 (1973), 199–221; Michael Aaron Dennis, "Accounting for Research: New Histories of Corporate Laboratories and the Social History of American Science," *Social Studies of Science* 17 (1987), 480–518; Charles E. Rosenberg, *No Other Gods: On Science and American Social Thought* (Baltimore and London, 1976), chaps. 8–12; Stuart Blume, Joske Bunders, Loet Leydesdorff, and Richard Whitley, eds., *The Social Direction of the Public Sciences: Causes and Consequences of Co-operation between Scientists and Non-scientific Groups*, Sociology of the Sciences Yearbook 11 (Dordrecht, 1987).

about its immediate practical use. By accommodating his initial disciplinary strategy and research program to changing conditions at home and abroad, Bjerknes accepted challenges that indeed proved fruitful.

The emergence of the Bergen meteorology illustrates new knowledge arising through changes in practice: discovery as a process. Hard facts were not waiting in nature to be uncovered. Bergen scientists constituted their new concepts and models by drawing upon analogy, metaphor, existing theory, and ad hoc construction. They transformed insights, speculations, and hypothetical entities into stable scientific "reality" by integrating these into a structure of meaning and by devising analytic techniques with which the constructs could be regularly reproduced.

The concepts and models were not the inevitable result of observation and theory. Indeed, after devising the method of indirect aerology and after establishing the existence of the occlusion process, Bergeron reanalyzed Jacob Bjerknes's weather maps from 14 and 15 August 1918 and showed that the cyclone used to support the original Bergen model was actually occluded and not a cyclone consisting of a clear warm sector bounded by cold air.[12] Observation itself can not only be said to be theory-laden but perhaps also practice-laden. Criteria for using weather played a significant role in how meteorologists and others saw weather phenomena; the meteorological requirements of postwar aviation, for instance, led meteorologists and weather observers to distinguish one hundred variations of weather. Theoretical and observational bases for a front concept predated the Bergen school, but only after practice changed did the conditions exist for constituting a front model that could be readily seen and used. In the case of the cyclone of 21 October 1921 with which this book opened, the Bergen-trained and the Danish meteorologists based their forecasts largely on the same data, but what the two groups saw differed considerably. Different perceptions arose from differing notions of practice.

In establishing new concepts, the Bergen meteorologists were also creating methods and analytic techniques that were not only to regulate how to see and interpret observations but also to provide the key to founding a school and spreading abroad its ideas, models, and research agenda. This pattern of basing a school on a particular method of analysis, which can be taught and which can produce results, is of course common in the history of science. Justus Liebig's legendary

12. Bergeron to Solberg, 9 June 1924, HSP.

chemical laboratory in midnineteenth-century Giessen, where students came to learn combustion analysis and left to take a range of jobs in universities and industry, enabled chemistry to advance professionally even though the theoretical underpinnings for the science were as yet open to disagreement and uncertainty.[13] For an ambitious scientist working on the geographic periphery with relatively few resources, devising a significant analytic or instrumental technique may provide a means not only to create a school but also to define the periphery as a center, as Bjerknes did in Bergen. Another example is The Svedberg, whose ultracentrifuge provided new methods for studying biochemical problems and enabled him to create an internationally significant school in Uppsala. But just as theories and ideas can be supplanted, so can methods and techniques upon which a school has been established.

Most of the world's meteorologists were using some form of the Bergen school methods by the close of World War II. More and more students and trainees in the growing number of academic meteorological departments were learning dynamic and practical meteorology from textbooks written by Bergen-trained and Bergen-inspired scientists. But in response both to technological advances in aerological instrumentation and numerical calculators and to theoretical concerns and challenges arising in part from new generations of aircraft, some atmospheric scientists were already preparing new methods of analysis based upon different theoretical perspectives; the Bergen school's hegemony was soon challenged. While many Bergen-school concepts have remained in meteorological discourse over the ensuing years, their meaning and significance have undergone repeated, even drastic, change. Although fronts and a cyclone model based on fronts are still used on television and newspaper weather maps as clear, simplified models for presenting weather situations, today's meteorologists work with considerably more sophisticated explanatory models, theories, and predictive methods.[14]

In contrast to most of twentieth-century meteorology, the Bergen meteorology, with its primary focus on the atmosphere's lower level, was an anomaly.[15] Most meteorologists at the time were focusing on

13. J. B. Morrell, "The Chemist Breeders: The Research Schools of Liebig and Thomas Thomson," *Ambix* 19 (1972), 1–45.

14. Rather amazingly, Norwegian television and the Oslo newspapers avoid fronts on their weather maps. To see an informative weather map in Oslo, one must consult Swedish television, the Stockholm morning papers, or a Bergen paper.

15. Bjerknes opposed efforts to consider the upper atmosphere as primary for understanding weather change; see Bjerknes to Wladimir Köppen, 9 April 1915, Köppen Papers, Graz University Library; Jacob Bjerknes, untitled MS, "I forbindelse med nogen cyklonundersøkelser . . . ," n.d. [Spring 1918], box I (1), JBP.

the upper levels of the atmosphere (here meaning upper troposphere and lower stratosphere) as the site for understanding weather change. Many, if not all, of the theoretical schemes that competed with the polar front meteorology considered the upper atmosphere crucial. After World War II, meteorologists not only began to use aerological data and upper-level analyses as much as surface data and analyses, but in increasing numbers they also began regarding lower-level atmospheric changes, such as the formation and movement of air masses, fronts, and cyclones, as consequences of events in the upper strata. It can perhaps be claimed that some of the contemporary alternatives to the Bergen meteorology have more in common with today's conceptions of atmospheric dynamics than did the polar front meteorology.[16] How then should the rise of the Bergen meteorology be understood?

The Bergen meteorology might best be regarded as a stopgap system that made the best of the political-economic and technological resources available to professional meteorology at the time. By focusing on processes and structures in the lower atmosphere, the Bergen school offered usable responses to the challenges aviation posed to meteorology. Its three-dimensional atmospheric models could be reproduced in practice using indirect aerology and the very limited direct aerological measurements then available. Theoretical schemes and statistical correlations relating upper-level patterns to weather phenomena in the lower atmosphere could provide marginal assistance for a meteorologist charged with preparing short-term forecasts for aviators.

In country after country, when flying became commercially and militarily significant, meteorologists were confronted with a dilemma: opportunity for major institutional growth was at hand, but what their science could do was not in perfect harmony with what it was being called on to deliver. Although atmospheric scientists could and did debate the problematic theoretical implications of the Bergen meteorology, for many of them, practical forecasting had to take precedence. Whether cyclones actually can be represented mathematically as waves on polar fronts that evolve into vortices troubled some meteorologists; for many others, that the polar front meteorology provided greater predictive certainty eventually outweighed the remaining theoretical difficulties.

In a science that could scarcely expect perfect agreement between theory and experience and that was often practiced in bureaucratic government settings, many issues necessarily figured in individual

16. This intriguing point was first brought to my attention by Arnt Eliassen.

and institutional conversion or resistance to the Bergen meteorology. Sometimes conversion was imposed from outside the science proper; Hermann Göring insisted the Bergen meteorology be used for securing the Luftwaffe. Such influential American scientists as Robert A. Millikan and Karl T. Compton worked through various organizational channels to institutionalize Bergen meteorology in America, a move they believed could help establish an academically oriented meteorological discipline useful to American aviation.[17]

To make sense of the Bergen era in meteorology's history, it will be necessary to understand more fully the national contexts in which the discipline developed. Detailed comparative national studies, especially of the German-Austrian schools, are called for. The vision and ambition of individual scientists cannot be ignored but must be understood within the political economy and culture of national science which set the conditions for career patterns and discipline building. In this broad historical context, perhaps some of the issues that began to come into play during and immediately after World War I will be seen more clearly.

Bjerknes's transformation of meteorology in Bergen was conditioned by two fundamental events. First, the Norwegian state expanded its role into commercial activities; the forecasting service provided a means of rationalizing the state's investment in weather-sensitive economic sectors. Second, regular commercial and military flying, for the security of which national governments assumed responsibility, required vastly expanded and improved weather services. Through Bjerknes's own quest to know and to succeed professionally, these issues played a constituent role in shaping a new meteorology. Throughout this book I have described the events from the perspective of Bjerknes making decisions against a changing background as he attempted to appropriate the weather. Perhaps a slight change in perspective—one that might be considered churlish— would show that Bjerknes's endeavors and subsequently those of other meteorological schools represent as well the manner by which aviation interests, but more fundamentally, nation-states, appropriated weather for their purposes.

17. See V. Bjerknes to Olle Arrhenius, 3 March 1936, copy, VBP, on events in Germany as related to him by Ludwig Weichmann; the Karl T. Compton Papers at the Massachusetts Institute of Technology and the Robert A. Millikan Papers at the California Institute of Technology are rich in correspondence on discipline building in American meteorology (I thank John Servos and Robert Kargon for bringing these collections to my attention); Richard P. Hallion, in *Legacy of Flight: The Guggenheim Contribution to American Aviation* (Seattle and London, 1977), provides some insight into the role of aviation interests in bringing Bergen methods to America.

Index

Library of Congress Cataloging-in-Publication Data

Friedman, Robert Marc.
  Appropriating the weather : Vilhelm Bjerknes and the construction of a modern
meteorology / Robert Marc Friedman.
     p. cm.
  Includes index.
  ISBN 0-8014-2062-8 (alk. paper)
     1. Bjerknes, V. (Vilhelm), 1862-   . 2. Meteorology—History. 3. Atmospheric
physics—History. 4. Meteorologists—Norway—Biography. I. Title.
QC858.B53F75 1989   551.5′092′4—dc19   88-47729